to the National Electrical Code®

ISBN 0-13-042209-6
90000
9 780130 422095

POCKET GUIDE TO THE NATIONAL ELECTRICAL CODE®

2002 EDITION

MARVIN J. FISCHER, P.E.

PH PTR

Prentice Hall PTR
Upper Saddle River, NJ 07458
www.phptr.com

Acquisitions Editor: Bernard Goodwin
Marketing Manager: Dan DePasquale
Editorial/production supervision: Pine Tree Composition
Manufacturing buyer: Alexis Heydt-Long
Cover Design Director: Jerry Votta
Cover Design: Karen Salzbach

Note: In reference to NFPA documents published prior to 2000, the paragraph numbering system has been updated to reflect NFPA's current style, which uses all dots instead of dashes and dots.

The publisher offers discounts on this book when ordered in bulk quantities.
For more information, contact:
Corporate Sales Department
Prentice Hall PTR
One Lake Street
Upper Saddle River, NJ 07458
Phone: 800–382–3419
FAX: 201–236–7141
E-mail: corpsales@prenhall.com

Printed in the United States of America

10 9 8 7 6 5 4 3 2 1

ISBN 0-13-042209-6

Pearson Education LTD
Pearson Education Australia PTY, Limited
Pearson Education Singapore, Pte. Ltd.
Pearson Education North Asia Ltd.
Pearson Education Canada, Ltd.
Pearson Educación de Mexico, S.A. de C.V.
Pearson Education—Japan
Pearson Education Malaysia, Pte. Ltd.

CONTENTS

4 Equipment for General Use 158

5 Special Occupancies 208

LIST OF TABLES

CHAPTER 9

ANNEX C

PREFACE

The *Pocket Guide to the National Electrical Code* is intended for use as a convenient reference guide for architects, engineers, electrical designers, electrical contractors, electricians, manufacturers, inspectors, teachers, students, and all those who use the code in their work. The reader should have a basic knowledge of electricity and the *National Electrical Code* or be using this book together with other instructional material in the study of electricity, electrical design, and construction. This book is not intended to replace the *National Electrical Code* but rather be used with it. It is intended to provide a reference book containing those topics in the *NEC*® that occur most frequently to the user in his or her daily work experience. The Code topics are discussed in plain understandable English. The size of this book is such that it can be carried at all times where it might prove inconvenient to carry a copy of the complete *National Electrical Code.*

The *NEC*® is probably the most referenced and widely distributed single electrical code in use today. It is adapted into law in most jurisdictions in the United States: town, village, city, county, and state. Most federal and state agencies adapt this code in whole or in part or refer to specific pertinent sections and paragraphs. Naturally, there are some exceptions. It is safe to say that if your work or profession in any way touches the electrical field, a thorough knowledge of the *National Electrical Code* may be quite important, if not an absolute necessity.

Let us examine for a moment who might be using the *NEC* and for what purpose. The reader will note that it is not uncommon today to find one person falling into more than one category. Those people in the design profession—architects, engineers, and electrical designers—must surely refer to the Code in their everyday work. Not only must the design conform to the Code theoretically, but practically the work must be capable of being installed in conformance with the Code so as to provide a safe installation. It should be noted that the *National Electrical Code* should be used as a guide for a minimum provision of an essentially safe installation. Adequate space must be provided, and proper wire size and insulation must be used for specific purposes. These are just a few of the many topics of concern for design professionals. All are aimed at providing safe installation.

Once a project is designed, it must be built. This is accomplished primarily by electrical contractors, their licensed electricians, and helpers. The *National Electrical Code* includes provisions for the proper installation and construction of electrical work. How to pull wire through conduit, number of wires in a conduit, color coding to be used, and where, how, and under what circumstances the various types of wire and cable may be used are just an example of the many areas of concern.

The supplies and equipment used in an electrical installation must meet minimum requirements that are enumerated in the Code. They must also meet a minimum set of standards, such as color coding and markings on wire and cable, thickness and gauge of materials, and voltage ratings. Hence the manufacturers are constantly referring to the *NEC* to ensure that their products meet the minimum standards set forth and therefore are saleable.

When work is finished, before being used, it must be inspected by the enforcing authority of the Code. It is at this point that many differences in interpretation emerge. Naturally, it would be nice if we could all come up with one meaning or intent for a given set of rules. Unfortunately, this is not always the case. The inspector must be as familiar with the Code as any other person using it and in some cases more so. He may become the final arbiter before a judge's decision.

Finally, let us not forget those who will take our place, the students and their teachers. The *National Electrical Code* becomes a useful instructional tool and therefore must be understood and used by the instructor and learned by the student.

It has been found that people using it need to refer to certain topics in the *National Electrical Code* more frequently than other topics. Some are of concern on a daily basis, whereas others may appear infrequently and then only when a special project is in progress. A copy of the complete *National Electrical Code* is a necessity and should always be available to the reader. However, a book containing a discussion of only those paragraphs or parts of paragraphs that occur most frequently, and one that is more convenient to carry than the entire Code, will obviously be an excellent aid for daily use. To add to the ease of use of this handy reference, it should be of a size that could easily be carried in a pocket and not take up space in a case. This *Pocket Guide to the National Electrical Code* was developed and written to fill that need and provide that convenience.

The topics and parts of articles, sections, and paragraphs of the *NEC* chosen for this book are those that the author felt would be most useful to the reader on a daily basis. Less frequently used parts of the *NEC* are noted by Article headings for reference purposes only. The material presented is not intended to appear as official interpretations or reproductions of the *NEC* by the author. Official or formal interpretations may be obtained only from the NFPA and through a definite procedure outlined in the Code. It is also recommended that the authorities having jurisdiction be consulted. This book refers to the 2002 edition of the *National Electrical Code*.

I would like to dedicate this book to my wife, *Ruth,* and children, *Michael* and *Cheryl, Cathi,* and *Wendy* and *Brad,* and grandchildren, *Brian, Julia, Jessica, Sarah, Candis, Michael,* and *Lexis* for their encouragement and patience during its writing.

Marvin J. Fischer

ARTICLE 80
ADMINISTRATION AND ENFORCEMENT

This is a new article that was added to the 2002 edition of the *National Electrical Code.* The introductory statement states that it is only informative unless the article is specifically adapted by a jurisdiction that adopts the *NEC.* The entire article reads like a law that might be in effect in any jurisdiction pertaining to the administration and enforcement requirements for electrical installations. The paragraphs that are contained in this article include: Scope, Definitions, Purpose, Adoption, Title, Application, Occupancy of Building or Structure, Authority, Electrical Board, Records and Reports, Permits and Approvals, Plans Review, Notice of Violations, Penalties, Connection to Electricity Supply and Inspector's Qualifications.

ARTICLE 90
INTRODUCTION

This article covers the purpose of the Code, the scope and arrangement of the Code, enforcement, mandatory rules, permissive rules and explanatory material, the method of obtaining formal interpretations, examination of equipment for safety, wiring planning and units of measurement. It should be noted that Section 90.1 states that the *National Electrical Code* is *not* intended to be used as a design specification. It contains only provisions for safety. Compliance with the Code does not ensure that the installation is adequate for good service or that it is convenient or efficient. These criteria are left to other qualified personnel such as electrical engineers or designers.

1

GENERAL

ARTICLE 100
DEFINITIONS

This article contains definitions. Other definitions may be found in sections for specific installations, such as Section 517.2 (Health Care Facilities) and Section 550.2 (Mobile Homes and Mobile Home Parks). There are many others.

ARTICLE 110
REQUIREMENTS FOR ELECTRICAL INSTALLATIONS

I. General

110.1 Scope

110.2 Approval

110.3 Examination, Identification, Installation, and Use of Equipment

110.4 Voltages

110.5 Conductors

Copper is to be used unless otherwise specified. Sizes given are for copper unless otherwise specified.

110.6 Conductor Sizes—Expressed in American Wire Gage (AWG) or in Circular Mils

110.7 Insulation Integrity

110.8 Wiring Methods

110.9 Interrupting Rating

110.10 Circuit Impedance and Other Characteristics

The entire electrical circuit, including all devices and wire, must be chosen and coordinated so that a fault will be cleared before extensive damage takes place.

110.11 Deteriorating Agents

110.12 Mechanical Execution of Work

110.13 Mounting and Cooling of Equipment

110.14 Electrical Connections

110.15 High-Leg Marking

110.16 Flash Protection

110.18 Arcing Parts

110.19 Light and Power from Railway Conductors

110.21 Marking

110.22 Identification of Disconnecting Means

110.23 Current Transformers

II. 600 Volts, Nominal, or Less

110.26 Working Space about Electrical Equipment (600 Volts, Nominal or Less)

(A) Clearances for three conditions are noted. Condition 1 consists of live exposed parts on one side and no live or grounded parts on the other side of the working space or guarded live parts (wood or other insulating material used) on both sides of the working space. Condition 2 consists of exposed live parts on one side and grounded parts on the other side. Condition 3 consists of exposed live parts on both sides unguarded with the operator between. For voltages to ground between 0 and 150 V, a minimum clear distance of 900 mm (3 ft.) is required in all cases. For voltages to ground between 151 and 600 V, Type 1 requires 900 mm (3 ft.) clearance. Type 2 requires 1 m (3½ ft.), and Type 3 requires 1.2 m (4 ft.). No working space is required behind assemblies where no renewable parts are on the back and all wiring is accessible from the front (such as panelboards). A minimum of 762 mm (30 in.) in width is required in front of electrical equipment as well as room for a 90° opening of doors.

(E) Headroom of 2 m (6½ ft.) is required above service equipment, motor control centers, panelboards, or switchboards, with the exception of dwelling units not in excess of 200 A. If the equipment exceeds 2 m (6½ ft.), headroom shall be at least the height of the equipment.

110.27 Guarding Live Parts

III. Over 600 Volts, Nominal

110.30 General

110.31 Enclosure for Electrical Installations

110.32 Work Space about Equipment

110.33 Entrance and Access to Work Space

110.34 Work Space and Guarding

110.36 Circuit Conductors

110.40 Temperature Limitations at Terminations

IV. Tunnel Installations Over 600 Volts, Nominal

110.51 General

110.52 Overcurrent Protection

110.53 Conductors

110.54 Bonding and Equipment Grounding Conductors

110.55 Transformers, Switches, and Electric Equipment

110.56 Energized Parts

110.57 Ventilation System Controls

110.58 Disconnecting Means

110.59 Enclosures

2

WIRING AND PROTECTION

ARTICLE 200
USE AND IDENTIFICATION OF GROUNDED CONDUCTORS

200.1 Scope

Article 200 covers the identification of terminals, grounded conductors in premises wiring systems, and identification of grounded conductors.

200.2 General

The wiring systems of all premises must have a grounded conductor that is identified as stated in Section 200.6. Exceptions to this are noted in Sections 210.10, 215.7, 250.21, 250.22, 250.162, 503.13, 517.63, 668.11, 668.21, and 690.41 Exception. When the grounded conductor is insulated, the insulation must be suitable for ungrounded conductors in the same circuits where the voltage is less than 1 kV or on impedance grounded neutral systems of 1 kV and over or for solidly grounded neutral systems of 1 kV or more [Section 250.184(a)], the insulation rating must be 600 V or more.

200.3 Connection to Grounded Systems

Premises Wiring. When the premises wiring system has a grounded conductor, it must be connected to a supply system which also has a grounded conductor.

200.6 Means of Identifying Grounded Conductors

(A) Sizes 6 AWG or Smaller. A white or gray covering or insulation or one with three continuous white stripes on other than green insulation is required on insulated grounded conductors of size 6 AWG or smaller. MI cable must have distinctive marking at the terminations. There are requirements for conductors in photovoltaic power systems, fixture wire, and aerial cable.

(B) Sizes Larger than 6 AWG. When the insulated grounded conductor is larger than 6 AWG, it can be identified the same as conductors smaller than 6 AWG or have white markings at its terminations. The marking must encircle the insulation or the conductor.

(C) Flexible Cords. When the grounded conductor is in a flexible cord it shall also have a white or gray insulation or covering or three continuous white stripes on other than green insulation or identified in acccordance with Section 400.22.

(D) Grounded Conductors of Different Systems.

(E) Grounded Conductors of Multiconductor Cables. There are two exceptions.

200.7 Use of Insulation of a White or Gray Color or with Three Continuous White Stripes

White or gray continuous coverings on a conductor, white or gray markings at terminations, or three continuous white stripes other than green insulation can be used only to identify the grounded conductor. There are noted conditions where white or gray may be used for other than the grounded conductor.

200.9 Means of Identification of Terminals

200.10 Identification of Terminals

When conductors are to be attached to more than "one side" of a circuit, they must have the device terminals identified. This is not necessary when the connection to the grounded conductor is evident, and in panelboards, and in devices rated over

30 A—other than polarized attachment plugs and receptacles. Polarized attachment plugs and receptacles must have the terminal for the grounded conductor marked white. Refer to Section 215.126 for equipment grounding terminals. The terminal for the grounded conductor in screw-shell devices is the one connected to the screw shell. If there is a lead connected to the screw shell, it must be white or gray.

200.11 Polarity of Connections

ARTICLE 210
BRANCH CIRCUITS

I. General Provisions

210.1 Scope

Article 210 applies to all branch circuits except motor loads. When a branch circuit supplies this type of load in combination with a motor load, both this article and Article 430 must be followed. When the branch circuit supplies only motor loads, Article 430 must be followed.

210.2 Other Articles for Specific-Purpose Branch Circuits

The rules for branch circuits in Table 210.2 amend or supplement this article as noted.

210.3 Rating

Branch circuits are rated according to the maximum permitted rating or setting of their overcurrent device. Generally, they are classified as 15, 20, 30, 40, and 50 A. There is an exception.

210.4 Multiwire Branch Circuits

Multiwire branch circuits are permitted. They can only supply line to neutral load, except if it supplies one piece of equipment or if all ungrounded conductors of the circuit are opened by the overcurrent device simultaneously. There is a dwelling unit restriction.

210.5 Identification for Branch Circuits

In general, the grounded conductor must be identified in accordance with Section 200.6. The equipment grounding conductor must be identified in accordance with Section 250.119.

TABLE 210.2 Specific-Purpose Branch Circuits

	Article	Section
Air-Conditioning and Refrigerating Equipment		440.6 440.31 440.32
Busways		368
Circuits and Equipment Operating at Less than 50 Volts	720	
Central Heating Equipment Other than Fixed Electric Space Heating Equipment		422.12
Class 1, Class 2, and Class 3 Remote-Control, Signaling, and Power-Limited Circuits	725	
Closed-Loop and Programmed Power Distribution Systems	780	
Cranes and Hoists		610.42
Electric Signs and Outline Lighting		600.6
Electric Welders	630	
Elevators, Dumbwaiters, Escalators, Moving Walks, Wheelchair Lifts, and Stairway Chair Lifts		620.61
Fire Alarm Systems	760	
Fixed Electric Heating Equipment for Pipelines and Vessels		427.4
Fixed Electric Space Heating Equipment		424.3
Fixed Outdoor Electric De-icing and Snow-Melting Equipment		426.4
Information Technology Equipment		645.5
Infared Lamp Industrial Heating Equipment		422.48, 424.3

	Article	Section
Induction and Dielectric Heating Equipment	665	
Marinas and Boatyards		555.19
Mobile Homes, Manufactured Homes, and Mobile Home Parks	550	
Motion Picture and Television Studios and Similar Locations	530	
Motors, Motor Circuits, and Controllers	430	
Pipe Organs		650.7
Recreational Vehicles and Recreational Vehicle Parks	551	
Sound-Recording and Similar Equipment		640.8
Switchboards and Panelboards		408.32
Theaters, Audience Areas of Motion Picture and Television Studios, and Similar Locations		520.41
		520.52
		520.62
X-Ray Equipment		660.2
		517.73

210.6 Branch-Circuit Voltage Limitations

(A) The voltage between conductors cannot be greater than 120 V for lighting fixtures and cord-and-plug connected loads of less than 1440 VA or less than ¼ hp in dwelling units, hotel and motel guest rooms, and the like.

(B) Circuits not exceeding 120 V to ground can supply auxiliary equipment of electric discharge lamps, utilization equipment, and terminals of lampholders within their voltage rating.

(C) Circuits between 120 and 277 V to ground can supply mogul-base screw-shell lampholders, lampholders used within their voltage rating, auxiliary equipment to electric discharge lamps, listed electric discharge fixtures, and utilization equipment. Additionally, these circuits can supply listed lampholders that are supplied at 120 V or less from a stepdown autotransformer which is part of the fixture and the outer shell terminal is connected to the grounded conductor of the branch circuit.

(D) 600 V between conductors.

(E) Over 600 Volts between conductors.

210.7 Branch Circuit Receptacle Requirements

(A) Receptacle Outlet Location. The location of receptacle outlets in branch circuits must be as noted in Part III of Article 210.

(B) Receptacle Requirements. Article 406 contains the specific requirements for receptacles.

(C) Multiple Branch Circuits. If more than one receptacle on the same yoke is supplied by more than one branch circuit then a means must be provided at the panelboard where the branch circuits originate to simultaneously disconnect the ungrounded conductors supplying those receptacles.

210.8 Ground-Fault Circuit-Interruptor Protection for Personnel

(A) Dwelling Units. All 125-V single-phase 15- and 20-A receptacles must have ground fault protection for personnel when installed in bathrooms, garages and at or below grade level parts of storage or work areas and similar use in accessory buildings (two exceptions), outdoors (one exception), crawl spaces at or below grade level, unfinished basements (two exceptions), kitchens where the receptacles are serving counter tops, and wet bar sinks where the receptacle is within 1.8 m

(6 ft.) of the edge of the sink and to serve the counter top. There are exceptions.

(B) Other than Dwelling Units. Ground-fault circuit interrupter receptacles are required in bathrooms, kitchens, and on roofs on all 125-V single-phase 15- and 20-A outlets. There is an exception.

210.9 Circuits Derived from Autotransformers

210.10 Ungrounded Conductors Tapped from Grounded Systems

210.11 Branch Circuits Required

Branch circuits have to be provided to supply the loads in accordance with Section 220.3. Also they are to be provided where required elsewhere in the Code, for dwelling units as specified in (C), and where not covered in Section 220.3.

(A) Number of branch circuits. The minimum number of branch circuits is determined by the total computed load and the rating or size of the circuits that are used. They have to supply the load and in no case can the load exceed the maximum specified in Section 220.4.

(B) Load Evenly Proportioned among Branch Circuits. When the load is calculated on a volt-ampere per square foot or square meter basis the wiring system up to and including the branch circuits panelboards has to be rated to carry the load. The load has to be evenly proportioned among the branch circuits. You only have to install the branch circuits and overcurrent devices for the connected load.

(C) Dwelling Units. Two or more 20 ampere small appliance branch circuits for outlets required by Section 210.52(b) are to be installed. One 20-ampere branch circuit is required for laundry [Section 210.52(F)] and bathroom receptacles. This circuit is to have no other outlets. These are in addition to other requirements. There is an exception for the bathroom.

210.12 Arc-Fault Circuit-Interrupter Protection

An arc fault circuit interrupter is a device which will offer protection from an arcing fault. This device is required in all branch circuits that supply 125-volt, single-phase, 15- and 20-ampere receptacles in dwelling unit bedrooms. It must provide protection for the entire branch circuit.

II. Branch-Circuit Ratings

210.19 Conductors—Minimum Ampacity and Size

(A) Branch Circuits No More Than 600 Volts.

(1) General. Generally, conductors must have a capacity not less than the maximum load. If a branch circuit supplies either continuous or continuous and noncontinuous loads, the ampere rating of the circuit must be at least equal to or greater than the noncontinuous load plus 125% of the continuous load. Various tables are supplied. A fine print note states that if the conductors are sized so that the maximum voltage drop on the branch circuit to the outlet is not greater than 3% of the total voltage drop, and for both feeder and branch circuit is not greater than 5%, a reasonable efficiency of operation will be provided.

(2) Multioutlet Branch Circuits. Conductors for receptacles for multioutlet branch circuits must be rated at least that of the branch circuit.

(3) Household Ranges and Cooking Appliances. Ratings cannot be less than the rating of the load. In cases of ranges of 8¾ kW or more, the minimum circuit rating is 40 A. Neutrals and conductors are permitted to be smaller in accordance with the Code. There are two exceptions.

(4) Other Loads. Except for cooking appliances and loads noted in Section 210.2, ratings must be sufficient for the loads supplied, but in no case can a conductor be smaller than 14 AWG. Exceptions are made for tap conductors and fixture wires and cords.

(B) Branch-Circuits Over 600 Volts. The ampacity of the conductors must be sized in accordance with Sections 310.15 and 310.60. Branch-circuit conductors over 600 volts have to be sized as follows:

(1) General. The ampacity of the conductors cannot be less than 125% of the load of equipment that will be used simultaneously. The designed potential load is to be used.

(2) Supervised Installations. A supervised installation is defined in the Code as the portions of a facility where two conditions are met. The first states that the conditions of the design are made under engineering supervision. The second states that the maintenance, monitoring, and servicing of the system is to be done with qualified personnel with documented training and experience in over 600 volt systems. Qualified personnel under engineering supervision can do branch-cricuit conductor sizing.

210.20 Overcurrent Protection

(A) Continuous and Noncontinuous Loads. If the circuit has continuous or a combination of continuous and noncontinuous loads then the rating of the overcurrent device has to be the sum of the noncontinuous and 125% of the continuous loads. There is an exception for devices listed for operation at 100% of rating.

(B) Conductor Properties. Conductors must be protected in accordance with Section 240.4.

(C) Equipment. Refer to applicable articles in Section 240.3.

(D) Outlet Devices. Do not exceed ratings or settings in Section 210.21.

210.21 Outlet Devices

Outlet devices must be rated at least that of the load.

(A) Lampholders connected to a branch circuit rated at more than 20 A must be a heavy-duty type.

(B) Receptacles. A single receptacle on an individual branch circuit can have a rating only equal to or larger than the rating of the individual branch circuit it is on except when installed in accordance with Section 430.81(C). The maximum load from a cord and plug connection where there is more than one receptacle on the circuit is shown in Table 210.21(B)(2). The ratings of receptacles where there is more than one on a circuit are shown in Table 210.21(B)(3). If the branch circuit rating is more than 50 A, then the receptacle rating must be at least as large as the branch circuit. The ampere rating of a range receptacle can be based on a single range as noted in Table 220.19. There are some exceptions.

TABLE 210.21(B)(2) Maximum Cord- and Plug-Connected Load to Receptacle

Circuit Rating (Amperes)	*Receptacle Rating (Amperes)*	*Maximum Load (Amperes)*
15 or 20	15	12
20	20	16
30	30	24

TABLE 210.21(B)(3) Receptacle Ratings for Various Size Circuits

Circuit Rating (Amperes)	*Receptacle Rating (Amperes)*
15	Not over 15
20	15 or 20
30	30
40	40 or 50
50	50

210.23 Permissible Loads

The load cannot exceed the branch circuit rating. Branch circuits that supply more than one receptacle or outlet follow the rules in (A) through (D) below and summarized in Section 210.24 and Table 210.24.

(A) 15 and 20 Ampere Branch Circuits. 15- and 20-A branch circuits can supply lighting loads, utilization equipment, or a combination. The rating of one cord- and plug-connected utilization equipment cannot be more than 80% of the branch-circuit ratings. Connected utilization equipment cannot be more than 50% of the branch-circuit rating where both connected and cord-and-plug loads are used.

(B) 30 Ampere Branch Circuit. A 30-A branch circuit can supply heavy-duty lampholders in fixed lighting units in all occupancies except dwelling units. They can also supply utilization equipment in all occupancies. One cord- and plug-connected utilization equipment cannot be more than 80% of the branch circuit rating.

(C) 40 and 50 Ampere Branch Circuits. A 40- and 50-A branch circuits supply the same lighting loads as the 30-A branch circuit, connected cooking appliances in any occupancy, and infrared heating units in occupancies dwelling units. They can supply fastened-in-place cooking appliances in dwelling units.

(D) Branch Circuits Larger Than 50 Amperes. If the branch circuit is larger than 50 A, it can only supply nonlighting loads.

210.24 Branch-Circuit Requirements—Summary

Table 210.24 provides a summary of branch-circuit requirements for circuits that contain two or more outlets or receptacles other than those required in Section 220.11(C)(1) and (2). Refer to 210.19, 210.20 and 210.21 for specifics to branch circuits.

210.25 Common Area Branch Circuits

Dwelling unit branch circuits can only supply circuits in or associated with that dwelling unit in multifamily dwellings. Branch circuits supplying loads such as lighting, alarms, and signals for common areas cannot be supplied from a dwelling unit panelboard.

III. Required Outlets

210.50 General

A cord connection is considered a receptacle outlet if it is supported by a permanently installed cord pendant. When it is known that a flexible cord will be used, a receptacle outlet should be installed. Appliance outlets for specific equipment are to be installed within 1.8 m (6 ft.) of the location of the appliance.

210.52 Dwelling Unit Receptacle Outlets

Requirements for 125 volt, 15 and 20 ampere receptacles are provided in this section.

(A) General Provisions. Receptacle outlets must be installed in dwelling units so that a receptacle outlet is not more than 1.8 m (6 ft.) from any point in a wall space along the floor line. This applies to wall spaces 600 mm (2 ft.) wide or larger, those containing fixed panels in exterior walls, and those containing fixed room dividers (i.e., railings or free standing bar-type counters). Sliding panels in exterior walls are excluded. A wall space is an unbroken wall along the floor (unbroken by doors, fireplaces, etc.). Each wall space 600 mm (2 ft.) wide or more is considered separately. Receptacles are to be spaced equally. Receptacles in floors and in lighting fixtures or appliances within cabinets or 1.7 m (5½ ft.) above the floor are not counted. The areas that this applies to are kitchen, family room, dining room, living room, parlor, library, den, sunroom, bedroom, recreation room, and similar types of rooms.

(B) Small Appliances. The required [Section 210.11(C)(1)] two or more 20-A small appliance branch circuits must serve all receptacles covered by Sections 210.52(A) and (C) in areas such as dining room, breakfast room, pantry, and kitchen, including refrigeration equipment. Exceptions are made for switched

TABLE 210.24 Summary of Branch-Circuit Requirements

Circuit Rating	*15 Amp*	*20 Amp*	*30 Amp*	*40 Amp*	*50 Amp*
Conductors: (min. size)					
Circuit wires[1]	14	12	10	8	6
Taps	14	14	14	12	12
Fixture wires and cords			Refer to Section 240.5		
Overcurrent Protection	*15 Amp*	*20 Amp*	*30 Amp*	*40 Amp*	*50 Amp*
Outlet devices:					
Lampholders permitted	Any type	Any type	Heavy duty	Heavy duty	Heavy duty
Receptacle rating[2]	15 Max. Amp	15 or 20 Amp	30 Amp	40 or 50 Amp	50 Amp

Maximum Load	15 Amp	20 Amp	30 Amp	40 Amp	50 Amp
Permissible load	Refer to Section 210.23(A)	Refer to Section 210.23(A)	Refer to Section 210.23(B)	Refer to Section 210.23(C)	Refer to Section 210.23(C)

[1] These gauges are for copper conductors.

[2] For receptacle rating of cord-connected electric-discharge lighting fixtures, see Section 410.30(C).

receptacles supplied from a general purpose branch circuit as described in Section 210.70(A), Exception 1 and refrigeration equipment can be supplied from a 15-A or greater individual branch circuit. No other outlets are allowed on the required two or more 20-A small appliance branch circuits except clocks and power for supplemental equipment on gas fired ranges, ovens, or counter-mounted cooking equipment. The receptacles in kitchens serving countertops must be supplied by at least two small appliance branch circuits. These small appliance branch circuits or additional small appliance branch circuits can supply outlets in the kitchen or other rooms as specified in Section 210.52(B)(1). No small appliance branch circuit can supply more than one kitchen.

(C) Countertops. A receptacle is to be installed in each wall countertop space wider than 300 mm (12 in.) in kitchens and dining areas. No point along the wall can be farther than 600 mm (24 in.) from a receptacle measured horizontally. If an island or peninsular countertop is 600 mm (24 in.) or more in one dimension and 300 mm (12 in.) or more in the other dimension it must have at least one receptacle. If a counter top is separated by a range, sink, or refrigerator each part is to be considered a separate counter top. No receptacle is to be more than 500 nm (20 in.) above the counter top. Receptacles are not to be installed face up. A receptacle which is covered by a fixed appliance or one in a dedicated space is not to be counted as a required outlet. There is an exception for the handicapped and a special island and peninsula condition.

(D) Bathrooms. A minimum of one wall receptacle is to be installed within 900 mm (3 ft.) of each basin. Receptacles cannot be mounted face up on countertops.

(E) Outdoor Outlets. A minimum of one outdoor outlet is to be installed outdoors for a one-family house. A two-family house must have at least one outdoor outlet for each unit that is at grade level. These receptacles are to be installed in the front and back. The receptacle outlet cannot be higher than 2.0 m (6½ ft.) above grade. Refer to Section 210.8(A)(3).

(F) Laundry Areas. A minimum of one outlet is to be installed in the laundry area. This is not required in multifamily dwelling units where laundry rooms are provided or laundry facilities are not individually installed.

(G) Basements and Garages. A minimum of one receptacle in addition to the one for the laundry is to be installed in the basement and garage with electricity of a one-family dwelling. Refer to Sections 210.8(A)(2) and (A)(5). If a portion of the basement is made a habitable room or rooms each unfinished portion must have a receptacle.

(H) Hallways. One receptacle must be installed in a hallway 3 m (10 ft.) or longer. The length of the hallway is measured along the centerline of the hall without going through a doorway.

210.60 Guest Rooms

Guest rooms in a hotel or motel must comply with Sections 210.52(A) and (D) unless otherwise required for permanently installed furniture. If a guest room meets the definition of a dwelling unit then it must comply with Section 210.52. There are requirements for permanent furniture layout.

210.62 Show Windows

A minimum of one receptacle is required for each 3.7 linear m (12 ft) or major fraction of a show window. It is to be measured at the maximum width and the receptacle to be installed above the window.

210.63 Heating, Air-Conditioning, and Refrigeration Equipment Outlet

210.70 Lighting Outlets Required

(A) Dwelling Units. Every habitable room, as well as attached garages, detached garages with electric power, hallways, stairways, and bathrooms, must have a minimum of one lighting outlet controlled by a wall switch. This is also required for illumination on the exterior of outdoor exits or entrances on grade level. A receptacle is acceptable in place of the lighting outlet in habitable rooms other than kitchens and bathrooms. Remote automatic or central control can be used in hallways, outdoor entrances, and stairways. When an attic, utility room, underfloor space, or basement is used for storage or equipment requiring service, it must have a minimum of one lighting outlet controlled at the point of entry. There are requirements for stair lighting. The lighting outlets can be controlled by an occupancy sensor if it is in addition to a wall switch or is at the location of the wall switch with a manual override.

(B) Guest Rooms. Guest rooms in hotels, motels, and similar installations must have a minimum of one lighting outlet or receptacle connected to a wall switch.

(C) Other Locations. Requirements for lighting in attics or underfloor spaces for heating, air-conditioning, refrigeration equipment, or other equipment needing servicing are noted here.

ARTICLE 215
FEEDERS

215.1 Scope

215.2 Minimum Rating and Size

The feeder conductors have to be sized to carry the load as computed in Article 220, Parts I, II, and III. The minimum size permitted before any adjustments is the total noncontinuous load and 125% of the continuous load. There is one exception where the assembly is rated for operation at 100% of its rating. If a three-wire feeder supplies more than two two-wire branch circuits or two or more three-wire branch circuits, it must be sized for at least 30 A. The same is true of a two-wire feeder supplying two or more two-wire branch circuits and two or more four-wire branch circuits supplied by a three-phase, four-wire feeder. If a feeder carries the total load of the service entrance conductors and the ampacity of the service entrance conductor is 55 A or less, the feeder must be sized at least the same. Feeders in mobile homes or individual dwelling units only have to be sized as large as the service entrance conductors. Reference is made to Section 310.15(B)(6) for sizing. There are also fine print notes which state that if the feeder conductors are sized to give a maximum voltage drop of 3% and the total voltage drop of feeders and branch circuits is not larger than 5%, operation should be reasonably efficient. There are requirements for feeders over 600 volts.

215.3 Overcurrent Protection

Refer to Article 240, Part I.

215.4 Feeders with a Common Neutral

215.5 Diagram of Feeders

215.6 Feeder Conductor Grounding Means

If equipment grounding conductors are required, refer to Section 250.134.

215.7 Ungrounded Conductors Tapped from Grounded Systems

Permitted with a switching device with a pole in each ungrounded conductor.

215.8 Means of Identifying Conductor with Higher Voltage to Ground

215.9 Ground-Fault Circuit-Interrupter Protection for Personnel

When a feeder supplies 15- and 20-A receptacle branch circuits, a ground-fault circuit interrupter may be installed in the feeder instead of the requirements of Section 210.8 and Article 527.

215.10 Ground-Fault Protection of Equipment

Ground-fault protection of equipment in accordance with Section 230.95 must be provided for a feeder in a solidly grounded wye system of 150 to 600 V to ground and a disconnect rated 1000 A or more. There are three exceptions, for a continuous industrial process where an orderly shutdown is required to prevent hazards, fire pumps, and if there is ground-fault protection equipment on the supply side of the feeder.

215.11 Circuits Derived from Autotransformers

ARTICLE 220
BRANCH-CIRCUIT, FEEDER, AND SERVICE CALCULATIONS

I. General

220.1 Scope

220.2 Computations

220.3 Computation of Branch-Circuit Loads

(A) Lighting Load for Specified Occupancies. Table 220.3(A) lists the minimum load required based on floor area, which is calculated using outside dimensions. Where an area cannot be used in dwelling units, garages, and open porches, it does not have to be included. It should be noted that these values are a minimum and may not provide for an adequate installation.

(B) Other Loads—All Occupancies. There is a list of requirements for all other loads. An outlet for an appliance or other specific load is to be calculated on the ampere rating of the appliance or load. The load for electric dryers is based on Section 220.18 and electric ranges and other cooking devices on Section 220.19. Motor loads are to be calculated in accordance with Sections 430.22, 430.24, and 440.6. Recesssed lighting fixtures are to be calculated based on the maximum volt-ampere rating of the equipment and lamps. Heavy-duty lampholders are calculated at a minimum of 600 VA. Sign and outdoor lighting is to be calculated at a minimum of 1200 VA for each branch circuit noted in Section 600.5(A). Show windows are calculated at 200 VA per linear foot

TABLE 220.3(A) General Lighting Loads by Occupancies

Type of Occupancy	Unit Load	
	Volt-Amperes per Sq. Meter	*Volt-Amperes per Sq. Ft.*
Armories and auditoriums	11	1
Banks	39[b]	3½[b]
Barber shops and beauty parlors	33	3
Churches	11	1
Clubs	22	2
Court rooms	22	2
Dwelling units[a]	33	3
Garages—commercial (storage)	6	½
Hospitals	22	2
Hotels and motels, including apartment houses without provisions for cooking by tenants[a]	22	2
Industrial commercial (loft) buildings	22	2
Lodge rooms	17	1½
Office buildings	39[b]	3½[b]
Restaurants	22	2
Schools	33	3
Stores	33	3
Warehouses (storage)	3	¼
In any of the above occupancies except one-family dwellings and individual dwelling units of two family and multi-family dwelling units:		
Assembly halls and auditoriums	11	1
Halls, corridors, closets, stairways	6	½
Storage spaces	3	¼

[a]See Section 220.3(B)(10).

[b]In addition a unit load of 11 volt-amperes/m² or 1 VA per square foot shall be included for general-purpose receptacle outlets when the actual number of general-purpose receptacle outlets is unknown.

of window or the unit load as noted elsewhere in this section. Multioutlet assemblies need only be counted as one outlet (180 VA) for each 1.5 m (5-ft.) length unless many appliances are used simultaneously. If appliances are used simultaneously, then each 1.5 m (1-ft.) length is calculated at 180 VA. The portion containing the receptacle outlets can be used for calculations. Receptacle outlets on one strap other than those covered by the calculations in Section 220.3(A) are calculated at 180 VA each. If a single piece of equipment has multiple receptacles, then each receptacle is to be calculated at not less than 90 VA. Receptacles noted in Sections 210.11(C)(1) and (2) are not bound by the previous two sentences. In dwelling occupancies all receptacle outlets as noted in Sections 210.11(C)(3), 210.52(E), 210.52(G), and 210.70(A) and (B) are included in the general lighting load calculations in Section 220.3(A). All other outlets are calculated at 180 VA.

(C) Loads for Additions to Existing Installations. Loads for additions greater than 46.5 square meters (500 square feet) to dwelling units are to be calculated in accordance with Section 220.3(A) and (B). Additions to circuits should be calculated as noted in Section 220.3(A) and (B). Loads for other than dwelling units are to be calculated as noted in Section 220.3(A) and (B).

220.4 Maximum Loads

The load should never exceed the rating of the branch circuit. Article 430 applies when only motor-operated loads are used. Article 440 is used for air-conditioning and refrigeration equipment. When more than one motor fastened in place is used, then you use 125% of the largest motor and the sum of the rest of the motor loads. Inductive lighting is calculated on the total ampere rating of the unit including ballasts, transformers, lamps, etc. Range load demand factors can be used in accordance with Table 220.19 including note 4.

II. Feeders and Services

220.10 General

Feeders must be able to supply the required load. They must be sized to supply the connected load with applied allowable demand factors.

220.11 General Lighting

To calculate the general lighting load, use the demand factors in Table 220.11.

TABLE 220.11 Lighting Load Demand Factors

Type of Occupancy	*Portion of Lighting Load to Which Demand Factor Applies (Volt-Amperes)*	*Demand Factor Percent*
Dwelling units	First 3000 or less at	100
	From 3001 to 120,000 at	35
	Remainder over 120,000 at	25
Hospitals[1]	First 50,000 or less at	40
	Remainder over 50,000 at	20
Hotels and motels—including apartment houses without provision for cooking by tenants[1]	First 20,000 or less at	50
	From 20,001 to 100,000 at	40
	Remainder over 100,000 at	30
Warehouses (storage)	First 12,500 or less at	100
	Remainder over 12,500 at	50
All others	Total volt-amperes	100

[1]The demand factors of this table shall not apply to the computed load of feeders or services supplying areas in hospitals, hotels, and motels where the entire lighting is likely to be used at one time, as in operating rooms, ballrooms, or dining rooms.

220.12 Show-Window and Track Lighting

Use at least 660 VA per linear meter 200 VA per linear foot for show windows and 150 VA for every 600 mm (2 ft.) for track lighting in other than dwelling units or hotels and motels. When multi-circuit track is used the load is to be considered divided equally between circuits.

220.13 Receptacle Loads—Nondwelling Units

Receptacle loads may either be lumped in with lighting loads and demand factors from Table 220.11 or figured separately and demand factors from Table 220.13 used. The load is to be figured at not more than 180 VA per receptacle in accordance with Section 220.3(B)(9) and Section 220.3(B)(8).

220.14 Motors

Calculate motor loads as directed by Sections 430.24, 430.25, and 430.26 and 440.6 for hermetic refrigerant motor compressors.

220.15 Fixed Electric Space Heating

Generally, the load is to be calculated at 100% of the connected load. In no case can the service or feeder current rating be less than that of the largest branch circuit supplied. If, however, the total load is less as a result of intermittent operation or diversity, the authorities can reduce the calculated load provided the conductors have a rating for the determined load.

220.16 Small Appliance and Laundry Loads—Dwelling Units

Use 1500 VA for each two-wire small-appliance and laundry branch circuit called for in Section 210.11(C)(1). This can be included in the general lighting load, and demand factors of Table 220.11 applied.

TABLE 220.13 Demand Factors for Nondwelling Receptacle Loads

Portion of Receptacle Load to Which Demand Factor Applies (Volt-Amperes)	*Demand Factor Percent*
First 10 kVA or less	100
Remainder over 10 kVA at	50

220.17 Appliance Load—Dwelling Units

A demand factor of 75% can be applied to nameplate ratings of four or more appliances in a dwelling supplied by the same feeder or service. This does not apply to electric ranges, clothes dryers, and space heating or air-conditioning equipment.

220.18 Electric Clothes Dryers—Dwelling Units

Use a load of 5000 VA or the nameplate rating, whichever is larger. Demand factors listed in Table 220.18 can be used. If two or more single-phase dryers are served by a three-phase four-wire circuit, the total load is calculated at twice the maximum number connected between two phases.

220.19 Electric Ranges and Other Cooking Appliances—Dwelling Unit(s)

The demand load for household electric ranges, wall-mounted ovens, counter-mounted cooking units, and other household cooking units larger than 1¾ kW are shown in Table 220.19. kVA and kW are considered equivalent for this section.

For single-phase ranges on a three-phase four-wire feeder or service, use twice the number of units between two phases.

TABLE 220.18 Demand Factors for Household Electric Clothes Dryers

Number of Dryers	*Demand Factor Percent*
1–4	100
5	85
6	75
7	65
8	60
9	55
10	50
11	47
12–22	[47–(number of dryers over 11)]%
23	35
24–42	[35–0.5(number of dryers over 23)]%
43 and over	25

TABLE 220.19 Demand Loads for Household Electric Ranges, Wall-Mounted Ovens, Counter-Mounted Cooking Units, and Other Household Cooking Appliances over 1¾ kW Rating

(Column C to be used in all cases except as otherwise permitted in Note 3.)

	Demand Factors (Percent) (See Notes)		*Column C Maximum Demand (kW) (See Notes)*
Number of Appliances	*Column A (Less Than 3½ kW Rating)*	*Column B (3½ kW to 8¾ kW Rating)*	*(Not over 12 kW Rating)*
1	80	80	8
2	75	65	11
3	70	55	14
4	66	50	17
5	62	45	20
6	59	43	21
7	56	40	22
8	53	36	23
9	51	35	24
10	49	34	25

TABLE 220.19 (Continued)

11	47	32	26
12	45	32	27
13	43	32	28
14	41	32	29
15	40	32	30
16	39	28	31
17	38	28	32
18	37	28	33
19	36	28	34
20	35	28	35
21	34	26	36
22	33	26	37
23	32	26	38
24	31	26	39
25	30	26	40
26-30	30	24	15 kW plus 1 kW for each range
31-40	30	22	
41-50	30	20	25 kW plus ¾ kW for each range
51-60	30	18	
61 and over	30	16	

Note 1. Over 12 kW through 27 kW ranges all of same rating. For ranges individually rated more than 12 kW but not more than 27 kW, the maximum demand in Column C shall be increased 5 percent for each additional kW of rating or major fraction thereof by which the rating of individual ranges exceeds 12 kW.

Note 2. Over $8^{3}/_{4}$ kW through 27 kW ranges of unequal ratings. For ranges individually rated more than $8^{3}/_{4}$ kW and of different ratings, but none exceeding 27 kW, an average value of rating shall be computed by adding together the ratings of all ranges to obtain the total connected load (using 12 kW for any range rated less than 12 kW) and dividing by the total number of ranges. Then the maximum demand in Column C shall be increased 5 percent for each kW or major fraction thereof by which this average value exceeds 12 kW.

Note 3. Over $1^{3}/_{4}$ kW through $8^{3}/_{4}$ kW. In lieu of the method provided in Column C, it shall be permissible to add the nameplate ratings of all household cooking appliances rated more than $1^{3}/_{4}$ kW but not more than $8^{3}/_{4}$ kW and multiply the sum by the demand factors specified in Column A or B for the given number of appliances. Where the rating of cooking appliances falls under both Column A and Column B, the demand factors for each column shall be applied to the appliances for that column, and the results added together.

Note 4. Branch-Circuit Load. It shall be permissible to compute the branch-circuit load for one range in accordance with Table 220.19. The branch-circuit load for one wall-mounted oven or one counter-mounted cooking unit shall be the nameplate rating of the appliance. The branch-circuit load for a counter-mounted cooking unit and not more than two wall-mounted ovens, all supplied from a single branch circuit and located in the same room, shall be computed by adding the nameplate rating of the individual appliances and treating this total as equivalent to one range.

Note 5. This table also applies to household cooking appliances rated over $1^{3}/_{4}$ kW and used in instructional programs.

Reference to the code should be made for five notes to Table 220.19 concerning ranges between 12 and 27 kW and between 1¾ and 8¾ kW for alternate allowable means of calculation.

220.20 Kitchen Equipment—Other Than Dwelling Units

This section applies to commercial cooking equipment, dishwasher booster heaters, water heaters, and other kitchen equipment. Demand factors shown in Table 220.20 can be used. They apply to thermostatic controlled or intermittently used kitchen equipment, not air conditioning, space heating or ventilating equipment. The feeder or service must always be larger than the sum of the two largest kitchen equipment loads.

220.21 Noncoincident Loads

If two loads on the same feeder or service will not be operating simultaneously, the smaller of the two can be omitted for calculation purposes.

220.22 Feeder or Service Neutral Load

The feeder or service neutral load is by definition the maximum unbalanced load between the neutral and one ungrounded conductor. In three-wire two-phase or five-wire two-phase systems the load is then multiplied by 140%. A factor of 70% is used in the case of household electric ranges, wall-mounted ovens, electric dryers, and counter-mounted cooking units. A factor of 70% may also be used for load more than 200 A in systems

TABLE 220.20 Demand Factors for Kitchen Equipment—Other Than Dwelling Unit(s)

Number of Units of Equipment	*Demand Factors (Percent)*
1	100
2	100
3	90
4	80
5	70
6 and over	65

with three-wire dc or single-phase ac; three-phase, four-wire; two-phase, three-wire; and two-phase, five-wire. Reduction of the grounded conductor capacity of a 3 wire circuit with two phase wires and neutral of a four-wire wye-connected three-phase system is not permitted.

III. Optional Calculations for Computing Feeder and Service Loads

220.30 Optional Calculation—Dwelling Unit

(A) Feeder and Service Load. This section can be used to calculate the load for a single three-wire, 120/240-V or 208Y/120-V feeder or service entrance with an ampacity of 100 A or larger. The neutral load can then be determined by Section 220.22. The load is the sum of (B) and (C) as follows:

(B) General Loads. These shall not be less than 100 percent of the first 10 kVA plus 40% of the remainder of the load listed as follows: each laundry and 20-A small appliance circuit at 1500 VA, for general lighting and receptacles use 33 VA per sq. m or 3 VA per sq. ft., the nameplate rating of all fastened in place appliances, all motors and low power factor loads use the nameplate ratings.

(C) Heating and Air Conditioning Load. Use the largest of the following.

1. 100% of the nameplate rating(s) of the air conditioning and cooling.
2. 100% of the nameplate ratings of the heat pump compressor and supplemental heat unless they both cannot operate at the same time.
3. 100% of the nameplate ratings of electric thermal storage and other heating systems where the usual load is expected to be continuous at the full nameplate value. Systems qualify under this selection shall not be figured under any other selection in (c).
4. 65% of the nameplate rating(s) of the central electric space heating including integral supplemental heating in heat pumps.
5. 65% of the nameplate rating(s) of electric space heating if less than four separately controlled units.
6. 40% of the nameplate rating(s) of electric space heating of four or more separately controlled units.

220.31 Optional Calculations for Additional Loads in Existing Dwellings

This section applies only to existing 120/240-V and 120/208Y-V three-wire services. It gives an alternate means of calculating loads. Refer to the section in the Code.

220.32 Optional Calculation—Multifamily Dwelling

This section permits an alternate means of calculation of feeder and service loads when each dwelling is supplied by a maximum of one feeder, has electric cooking, and has electric space heating or air conditioning. The alternate calculation is based on the demand factors shown in Table 220.32. These demand factors apply to the small-appliance branch-circuit load, 33 VA per sq. m or 3 VA per square foot general load, the nameplate rating of all appliances and motors, and air-conditioning or space heating loads. House loads are calculated in accordance with Part II of this article and are in addition to the dwelling unit loads calculated by Table 220.32. There is a listing of the connected loads which apply to the table.

220.33 Optional Calculation—Two Dwelling Units

220.34 Optional Method—Schools

Refer to Table 220.34.

220.35 Optional Calculations for Determining Existing Loads

Actual demand data can be used if the maximum readings are known for at least 1 year, the existing demand multiplied by 125% and the new loads are not larger than the wire capacity, and proper feeder and service overcurrent protection are provided, in accordance with Sections 240.3 and 230.90, respectively. There is an exception if the maximum demand is not available for 1 year.

220.36 Optional Calculations—New Restaurants

Table 220.36 can be used to calculate a service load or a feeder for a new restaurant if the feeder serves the total load. The service-entrance overload protection must be in accordance with Sections 230.90 and 240.3. The feeder conductors do not have to be larger than the service-entrance conductors. If this optional calculation is used, the neutral load of service or feeder conduc-

TABLE 220.32 Optional Calculation—Demand Factors for Three or More Multifamily Dwelling Units

Number of Dwelling Units	*Demand Factor (Percent)*
3-5	45
6-7	44
8-10	43
11	42
12-13	41
14-15	40
16-17	39
18-20	38
21	37
22-23	36
24-25	35
26-27	34
28-30	33
31	32
32-33	31
34-36	30
37-38	29
39-42	28
43-45	27
46-50	26
51-55	25
56-61	24
62 and over	23

tors can be determined by Section 220.22. Table 220.36 indicates demand factors for all-electric and not-all-electric loads.

IV. Method for Computing Farm Loads

220.40 Farm Loads—Buildings and Other Loads

220.41 Farm Loads—Total

Note: Examples for the calculations referred to in Article 220 are shown in Annex D of the *NEC*®. They are shown in the same location in this book for correlation.

TABLE 220.34 Optional Method—Demand Factors for Feeders and Service-Entrance Conductors for Schools

Connected Load		*Demand Factors*
VA/m2	*VA/ft2*	*(Percent)*
The first 33 VA/m^2 at Plus,	The first 3 VA/ft^2 at Plus,	100
Over 33 to 220 VA/m^2 at Plus,	Over 3 to 20 VA/ft^2 at Plus,	75
Remainder over 220 VA/m^2 at	Remainder over 20 VA/ft^2 at	25

ARTICLE 225
OUTSIDE BRANCH CIRCUITS AND FEEDERS

225.1 Scope

This article covers feeders and branch circuits located outside on premises between buildings or poles, or for equipment on the outside of buildings or poles.

225.2 Other Articles

Table 225.2 refers to other articles in the Code which are pertinent to outside branch circuits and feeders.

I. General

225.3 Calculation of Loads 6 Volts, Nominal, or Less

The same as in Article 220.

225.4 Conductor Covering

Conductors must be insulated or covered within 3 m (10 ft.) of a building. Conductors in raceways or cables except MI must have thermoplastic or rubber insulation. An exception is made for equipment grounding conductors and grounded circuit conductors. Section 310.8 governs wet locations.

TABLE 225.2 Other Articles

	Article
Branch Circuits	210
Class 1, Class 2, and Class 3 Remote-Control, Signaling, and Power-Limited Circuits	725
Communications Circuits	800
Community Antenna Television and Radio Distribution Systems	820
Conductors for General Wiring	310
Electrically Driven or Controlled Irrigation Machines	675
Electric Signs and Outline Lighting	600
Feeders	215
Fire Alarm Systems	760
Fixed Outdoor Electric De-icing and Snow-Melting Equipment	426
Floating Buildings	553
Grounding	250
Hazardous (Classified) Locations	500
Hazardous (Classified) Locations, Specific	510
Marinas and Boatyards	555
Messenger Supported Wiring	396
Open Wiring on Insulators	398
Over 600 Volts, General	490
Overcurrent Protection	240
Radio and Television Equipment	810
Services	230
Solar Photovoltaic Systems	690
Swimming Pools, Fountains, and Similar Installations	680
Use and Identification of Grounded Conductors	200

225.5 Size of Conductors 600 Volts, Nominal, or Less

Size in accordance with Section 310.15. Determine loads in accordance with Section 220.3 and Part II of Article 220.

225.6 Conductor Size and Support

(A) Overhead Spans

600 V or less, up to 15 m (50 ft.): 10 AWG copper, 8 AWG aluminum except where supported by a messenger wire

600 V or less, over 15 m (50 ft.): 8 AWG copper, 6 AWG aluminum except where supported by a messenger wire
More than 600 V: 6 AWG copper, 4 AWG aluminum for open conductors; 8 AWG copper, 6 AWG aluminum for cable

(B) Festoon Lighting

Use 12 AWG or Larger for festoon lighting except where messenger wires support them, over 12 m (40-ft.) span require messenger wire support.

225.7 Lighting Equipment Installed Outdoors

225.9 Overcurrent Protection

Refer to Section 210.20 and Article 240.

225.10 Wiring on Buildings

225.11 Circuit Exits and Entrances

Refer to Sections 230.52, and 230.54.

225.12 Open Conductor Supports

225.14 Open Conductor Spacings

225.15 Supports Over Buildings

Refer to Section 230.29.

225.16 Point of Attachment to Buildings

Refer to Section 230.26.

225.17 Means of Attachment to Buildings

Refer to Section 230.27.

225.18 Clearance from Ground

Open conductors and open multiconductor cables not over 600 V must conform to the following:

3 m (10 ft.): above finished grade: 150 V to ground
3.7 m (12 ft.): over residential driveways and commercial areas not subject to truck traffic: 300 V to ground limitation
4.5 m (15 ft.): same as 3.7 m (12 ft.) above: exceeding 300 V to ground
5.5 m (18 ft.): above streets, driveways, etc. subject to truck traffic as well as land used by vehicles such as forest, grazing, etc.

225.19 Clearances from Buildings for Conductors of Not over 600 Volts, Nominal

(A) Above Roofs. Not less than 2.5 m (8 ft.) vertical clearance from the roof surface maintained for a distance of not less than 900 mm (3 ft.) in all directions from the edge of the roof except the final span to the attachment to the building. If subject to pedestrian or vehicular traffic, vertical clearances should be in accordance with Section 225.18. Other exceptions are as follows:

900 mm (3 ft.): voltage between conductors not more than 300 V and roof slope is maximum of 100 mm (4 in.) in 300 mm (12 in.)
450 mm (18 in).: voltage between conductors not more than 300 V, not more than 1.8 m (6 ft.) of conductor, 1.2 m (4 ft.) horizontally, do not pass over an overhang and are terminated in approved support or raceway through the roof
900 mm (3 ft.): clearance does not apply to the final conductor span attached to the roof

(B) From Nonbuilding or Nonbridge Structures. Clearances in all directions to be a minimum of 900 mm (3 ft.).

(C) Horizontal Clearances. Not less than 900 mm (3 ft.).

(D) Final Spans. Those may be attached to a building but must clear operable windows, fire escapes, and the like by at least 900 mm (3 ft.). If above the top level of a window then it can be reduced to less than 900 mm (3 ft.). There is additional information concerning openings in commercial buildings and vertical clearances.

(E) Zone for Fire Ladders. Allow for fire ladders—at least a 1.8 m (6-ft.)-wide space.

225.20 Mechanical Protection of Conductors

Refer to Section 230.50.

225.21 Multiconductor Cables on Exterior Surfaces of Buildings

Refer to Section 230.51.

225.22 Raceways on Exterior Surfaces of Buildings, or Other Structures

225.24 Outdoor Lampholders

225.25 Location of Outdoor Lampholders

225.26 Vegetation as Support

II. More Than One Building or Other Source

225.30 Number of Supplies

Where you have more than one building under one management and on the same property you can only have one feeder or branch circuit supplying each of the additional buildings. The exceptions are noted in (A) through (E).

(A) Special Conditions. You have additional feeders for fire pumps, emergency systems, legally required standby systems, parallel power production systems, and optional standby systems.

(B) Special Occupancies. In a single building where the load is too large for one feeder or branch circuit or in multi-occupancy buildings where there is no space for supplies accessible to all occupants then with special permission you can have additional feeders or branch circuits.

(C) Capacity Requirements. If the load requirements are more than 2000 A and the voltage is not more than 600 V, more than one supply is allowed.

(D) Different Characteristics. More than one supply is allowed if different characteristics, i.e., voltage, frequency, etc. are required.

(E) Documented Switch Procedures. You can have more than one supply if they are under one management and there are documented safe disconnecting procedures.

225.31 Disconnecting Means

Must be provided.

225.32 Location

It must be located at the poiint where the wire passes through the building either inside or outside. It must be readily accessible. You are allowed to use the requirements of Section 230.6. There are four exceptions.

225.33 Maximum Number of Disconnects

The provision for disconnecting each supply can consist of no more than six switches or six circuit breakers. Single-pole units can be used on multiwire circuits, one pole for each ungrounded conductor. They must be tied together as a multipole disconnect and will then be counted as one disconnect. An exception is made for the control circuit of a ground-fault protection system or the control circuit of the power operated supply disconnecting means.

225.34 Grouping of Disconnects

All disconnects are to be grouped together. An exception is made for fire pumps. The disconnects for fire pumps, emergency legally required standby or optional standby systems must be located away from the normal supply.

225.35 Access to Occupants

Must be accessible to the occupants except where there is a single management with continuous supervision.

225.36 Suitable for Service Equipment

225.37 Identification

225.38 Disconnect Construction

225.39 Rating of Disconnect

In general, the feeder or branch circuit disconnect must be sized to carry the calculated load. There are special cases where a minimum size is required. They are:

15 A: a single branch-circuit load.
30 A: up to two, two-wire branch circuits
100 A: three-wire: one-family dwelling with six or more two-wire branch circuits or calculated load is 10 kW or more.
60 A: all others.

225.40 Access to Overcurrent Protective Devices

III. Over 600 Volts

225.50 Sizing of Conductors

225.51 Isolating Switches

225.52 Locations

225.53 Type

225.60 Clearances over Roadways, Walkways, Rail, Water, and Open Land

225.61 Clearances over Buildings and Other Structures

ARTICLE 230
SERVICES

230.1 Scope

This article describes requirements for service conductors and equipment for control and protection of services. Figure 230.1 describes where in Article 230 different parts of the service are described.

I. General

230.2 Number of Services

A building normally can have only one service. If more than one is permitted as a result of the exceptions provided, then a directory of all services must be placed at each service. For the purposes of Section 230.40, Exception 2, underground conductors sized at least 1/0 are considered one service if they are connected at the supply but not at the load. More than one service is permitted as follows:

(A) Special Conditions

(1) Fire pumps if required.

(2) Emergency systems.

(3) Legally required standby systems.

(4) Optional standby systems.

(5) Parallel power production systems.

(B) Special Occupancies

(1) In multiple-occupancy buildings where there is no room for each tenant to have access to the service equipment, with permission, more than one set of service conductors can be tapped from one service drop or service lateral.

(2) If a single building is too large then with special permission, more than one service may be installed.

(C) Capacity Requirements

(1) If the capacity requirements are larger than 2000 amperes at 600 volts or less.

(2) If a single-phase installation were larger than the utility would normally serve through one service.

(3) With special permission.

(D) Different Characteristics

More than one service is permitted if there are different rate schedules, voltages, phases, frequencies or uses.

(E) Identification

If a building is supplied by more than one service, feeder, or set of branch circuits, a permanent directory or plaque has to be installed at the disconnect location.

230.3 One Building or Other Structure Not to Be Supplied through Another

Service conductors cannot pass through another building.

230.6 Conductors Considered Outside of Building

If conductors have a cover of at least 50 mm (2 in.) of concrete under a building or are installed in a raceway encased in at least 50 mm (2 in.) of concrete or are installed in a vault as per Article 450 Part III, or installed in concrete and under at least 450 mm (18 inches) of earth under a building they are treated as being outside the building.

230.7 Other Conductors in Raceway or Cable

230.8 Raceway Seal

230.9 Clearance from Building Openings

If installed as open conductors or multiconductor cable without an outer jacket, clearance of at least 900 mm (3 ft.) is required from operable windows, doors, porches, fire escapes, and the like. Overhead service conductors are not permitted underneath openings through which materials can be moved or obstructing these openings. Vertical clearances under certain conditions are to be maintained in accordance with Section 230.24(B).

230.10 Vegetation as Support Not Permitted

II. Overhead Service—Drop Conductors

230.22 Insulation or Covering

Insulation or covering is required on individual conductors.

230.23 Size and Rating

Conductors should have the capacity required for the load in accordance with Article 220 and have enough mechanical strength. The minimum size should be 8 AWG copper or 6 AWG aluminum or copper-clad aluminum. Grounded conductors shall be as required in Section 250.24(B). There is an exception.

230.24 Clearances

Service-drop conductors must not be accessible and comply with the following:

(A) 2.5 m (8 ft.) vertical clearance maintained above the roof with 900 mm (3 ft.) from the edge of the roof is generally the clearance required above roofs with a nominal voltage of 600 V except where the service drop is attached to the side of the building. This is reduced to 900 mm (3 ft.) with a voltage of 300 V and a roof slope of 100 mm (4 in.) in 300 mm (12 in.) or more. It is also reduced to 450 mm (18 in.) above the overhanging portion with a nominal voltage of 300 V if only up to 1.8 m (6 ft.) of service drop conductors, 1.2 m (4 ft.) horizontally, pass over the roof and the conductors end on a support or raceway. Where there is traffic above a roof surface use requirements of Section 230.24(B).

(B) The clearances above the ground vary for conductors with a nominal voltage of 600 V. They are:

3.0 m (10 ft.): voltage of 150 V to ground, service entrance to buildings accessible to pedestrians.

3.7 m (12 ft.): over residential property and no truck traffic commercial areas with voltage to ground not in excess of 300.

4.5 m (15 ft.): at areas listed in 3.7 m (12 ft.) classification when the voltage exceeds 300 V to ground.

5.5 m (18 ft.): over public streets. Private property is included if traveled with commercial vehicles.

(C) Refer to Section 230.9 for clearances from building openings.

(D) Refer to Section 680.8 for clearances from swimming pools.

230.26 Point of Attachment

A minimum clearance of 3.0 m (10 ft.) is required. Compliance with Section 230.24 is required.

230.27 Means of Attachment

230.28 Service Masts as Support

230.29 Supports over Buildings

III. Underground Service—Lateral Conductors

230.30 Insulation

Generally, they are insulated with the voltage determining the insulation. There are some exceptions for grounded conductors of bare wire and aluminum or copper-clad aluminum cable assemblies.

230.31 Size and Rating

The conductors must be large enough to carry the load in accordance with Article 220. No smaller than 8 AWG for copper and 6 AWG for aluminum or copper-clad aluminum. An exception is made where the service supplies only a single load on a single branch circuit. The grounded conductor must not be sized less than required in Section 250.24(B).

230.32 Protection against Damage

230.33 Spliced Conductors

IV. Service-Entrance Conductors

230.40 Number of Service-Entrance Conductor Sets

Generally, a service drop or lateral can only supply one set of service conductors. However, one set of service conductors can supply each occupancy or a group of occupancies in a multiple-occupancy building. A second exception for separate enclosures is noted. A third exception is noted for single family or a separate structure. There are two additional exceptions.

230.41 Insulation of Service-Entrance Conductors

Generally, if the conductors are on or enter a building, they should be insulated. Exceptions are made for grounded conductors of bare copper wire and aluminum, copper-clad aluminum cable assemblies in special cases, copper for direct burial where the copper is all right for the soil, and aluminum or copper-clad aluminum part of an assembly rated for direct burial.

230.42 Minimum Size and Rating

(A) General. The ampacity of the conductors is to be either the sum of the noncontinuous load and 125% of the continuous load or the sum of the noncontinuous and continuous loads if the overcurrent devices are rated for operation at 100% of their rating. The loads are calculated in accordance with Article 220.

The ratings are determined in accordance with Section 310.15. For busways use the listing or rating of the busway.

(B) Ungrounded Conductors. Use at least the rating of the disconnecting means as specified in Section 230.79(A) through (D).

(C) Grounded Conductors. Size in accordance with Section 250.24(B).

230.43 Wiring Methods for 600 Volts, Nominal, or Less

Only the following methods are permitted: Open wiring on insulators; type IGS cable; rigid metal conduit; intermediate metal conduit; EMT; ENT; service entrance cables; wireways; busways; auxiliary gutters; rigid NMC; cablebus; MC cable; mineral-insulated, metal-sheathed cable; flexible metal conduit or liquidtight flexible metal conduit with certain restrictions; or liquidtight flexible nonmetallic conduit.

230.44 Cable Trays

To support cables, a cable tray system may be used.

230.46 Spliced Conductors

230.49 Protection against Damage—Underground

230.50 Protection of Open Conductors and Cables against Damage—Above Ground

230.51 Mounting Supports

230.52 Individual Conductors Entering Buildings or Other Structures

230.53 Raceways to Drain

230.54 Overhead Service Locations

Connections have to be made raintight with a raintight head or with a gooseneck. The heads or goosenecks must be above the point where the conductors are attached to the building and must be securely attached. Drip loops must be constructed to prevent entrance of moisture. There are other requirements.

230.56 Service Conductor with the Higher Voltage to Ground

V. Service Equipment—General

230.62 Service Equipment—Enclosed or Guarded

Energized parts must be enclosed or installed in switchboard or similar equipment as stated in Sections 110.18 and 110.27.

230.66 Marking

Service equipment must be marked for use as service equipment.

VI. Service Equipment—Disconnecting Means

230.70 General

There must be a way to disconnect all conductors in the building from the service-entrance conductors. The disconnect must be readily accessible either outside the building or inside the building nearest the point of entrance of the service conductors. It must have proper markings and be suitable for its use. Service disconnects cannot be installed in bathrooms.

230.71 Maximum Number of Disconnects

The provision for disconnecting each service (Section 230.2) or the service entrance conductors (Section 230.40, Exception Nos. 1, 3, 4, or 5) can consist of no more than six switches or six circuit breakers or combinations. Single-pole units can be used for multiwire circuits, one pole for each ungrounded conductor. They must be tied together as a multipole disconnect and will then be counted as one disconnect. An exception is made for the control circuit of a ground-fault protection system, power monitoring equipment, transient voltage surge suppression, or power-operable service disconnecting means.

230.72 Grouping of Disconnects

All disconnects are to be grouped together. Exceptions are for services mentioned in Section 230.71 and when a switch is used for a fire water pump. The exceptions should be located far enough away from the service disconnects to minimize simultaneous interruption. Where multiple occupancies are present, each occupant must have access to his or her own disconnect. There are two exceptions to this latter requirement for certain management conditions.

230.74 Simultaneous Openings of Poles

230.75 Disconnection of Grounded Conductor

230.76 Manually or Power Operable

230.77 Indicating

230.79 Rating of Disconnect

In general, the service disconnect must be sized to carry the calculated load. There are special cases where a minimum size is required. They are:

15 A: a single branch-circuit load
30 A: up to two, two-wire branch circuits
100 A: three-wire: one-family dwelling
60 A: all other cases

230.80 Combined Ratings of Disconnects

If more than one service switch is used, their combined rating must be at least the rating shown in Section 230.79.

230.81 Connection to Terminals

230.82 Equipment Connected to the Supply Side of Service Disconnect

Among the pieces of equipment allowed to be connected to the supply side of a service switch are: current-limiting devices, meters, instrument transformers, surge protective devices, fire and sprinkler alarms, fire pump equipment, and interconnected electric production sources. Some of these items require separate protection and disconnects. Refer to the *NEC*® for a complete description and list.

VII. Service Equipment—Overcurrent Protection

230.90 Where Required

(A) Every ungrounded conductor must have an overcurrent device. It must normally not be set higher than the rating of the conductors. There are five exceptions to this and they include: motor starting current; fuses and circuit breakers conforming to Section 240.3(B) or (C) and Section 240.6; six fuses or circuit breakers is the limit for consideration as a service overcurrent device; fire pumps under certain conditions; when permitted by Section 310.15(B)(6) for 120/240 volt, three-wire, single-phase dwelling services.

(B) No overcurrent device is to be placed in a grounded conductor.

230.91 Location of Overload Protection

The service overcurrent device must be part of the service disconnect or be next to it.

230.92 Locked Service Overcurrent Devices

230.93 Protection of Specific Circuits

230.94 Relative Location of Overcurrent Devices and Other Service Equipment

Some of the items allowed ahead of the overcurrent device are: service switch, surge protectors, protective signaling systems, fire pump, meters, and power-operable service equipment controls.

230.95 Ground-Fault Protection of Equipment

This is required for all services that are 150 V to ground up to 600 V phase to phase, solidly grounded wye, with service switches 1000 A or more. The rest of this section contains information about the setting of these devices, the use of switches and fuses, and performance testing. An exception is allowed for continuous industrial processes where a nonorderly shutdown would be hazardous and for fire pumps.

VIII. Services Exceeding 600 Volts, Nominal

230.200 General

230.202 Service-Entrance Conductors

230.204 Isolating Switches

230.205 Disconnecting Means

230.206 Overcurrent Devices as Disconnecting Means

230.208 Protection Requirements

230.209 Surge Arresters (Lightning Arresters)

230.210 Service Equipment—General Provisions

230.211 Metal-Enclosed Switchgear

230.212 Services over 35,000 Volts

ARTICLE 240
OVERCURRENT PROTECTION

I. General

240.1 Scope

Parts I through VII give requirements for circuits of up to 600 V. Part VIII covers supervised industrial installations up to 600 V. Part IX covers over 600 V.

240.2 Definitions

240.3 Other Articles

Table 240.3 refers to other articles in the Code which pertain to overcurrent profection for specific equipment and systems.

240.4 Protection of Conductors

For other than flexible cords and fixture wires, use ampacities specified in Section 310.15 when calculating overcurrent protection. There are a number of cases where this rule does not necessarily hold true with special requirements. Some of them are: power loss hazard; devices rated 800 amperes or less; devices rated over 800 amperes; tap conductors; motor-operated appliance circuit conductors; motor and motor control circuit conductors; phase converter supply conductors; air conditioning and refrigeration equipment circuit conductors; transformer secondary conductors; capacitor circuit conductors; electric welder circuit conductors; remote control signalling and power limited circuit conductors; and fire protective alarm system circuit conductors. Refer to the *NEC*® for a complete list.

240.5 Protection of Flexible Cords, Flexible Cables, and Fixture Wires

240.6 Standard Ampere Ratings

The standard ratings for fuses and inverse time circuit breakers are: 15, 20, 25, 30, 35, 40, 45, 50, 60, 70, 80, 90, 100, 110, 125, 150, 175, 200, 225, 250, 300, 350, 400, 450, 500, 600, 700, 800, 1000, 1200, 1600, 2000, 2500, 3000, 4000, 5000, and 6000 A. Fuses are also rated at 1, 3, 6, 10, and 601 A. There is a requirement for circuit breakers having external adjustable means and for restricted access adjustable trip circuit breakers.

240.8 Fuses or Circuit Breakers in Parallel

240.9 Thermal Devices

240.10 Supplementary Overcurrent Protection

240.12 Electrical System Coordination

If an orderly shutdown is required to reduce hazards to equipment and personnel, the following method can be used; coordinated short-circuit protection and/or overload indication.

TABLE 240.3 Other Articles

	Article
Air-Conditioning and Refrigerating Equipment	440
Appliances	422
Audio Signal Processing Amplification, and Reproduction Equipment	640
Branch Circuits	210
Busways	368
Capacitors	460
Class 1, Class 2, and Class 3 Remote-Control, Signaling, and Power-Limited Circuits	725
Closed-Loop and Programmed Power Distribution System	780
Cranes and Hoists	610
Electric Signs and Outline Lighting	600
Electric Welders	630
Electrolytic Cells	668
Elevators, Dumbwaiters, Escalators, Moving Walks, Wheel Chair Lifts, and Stairway Chair Lifts	620
Emergency Systems	700
Fire Alarm Signaling Systems	760
Fire Pumps	695
Fixed Electric Heating Equipment for Pipelines and Vessels	427
Fixed Electric Space Heating Equipment	424
Fixed Outdoor Electric De-icing and Snow-Melting Equipment	426
Generators	445
Health Care Facilities	517
Induction and Dielectric Heating Equipment	665
Industrial Machinery	670
Luminaires (lighting fixtures), Lampholders, Lamps, and Receptacles	410
Motion Picture and Television Studios and Similar Locations	530
Motors, Motor Circuits, and Controllers	430
Phase Converters	455
Pipe Organs	650
Places of Assembly	518

TABLE 240.3 Other Articles (continued)

	Article
Services	230
Solar Photovoltaic Systems	690
Switchboards and Panelboards	408
Theaters, Audience Areas of Motion Picture and Television Studios, and Similar Locations	520
Transformers and Transformer Vaults	450
X-Ray Equipment	660

240.13 Ground-Fault Protection of Equipment

For solidly grounded wye systems that are more than 150 V to ground and not exceeding 600 V phase to phase, ground-fault protection of equipment must be provided for each main disconnecting means rated 1000 A or more, in accordance with Section 230.95. The three exceptions are where an orderly shutdown in an industrial process is required, for fire pumps, and if the disconnect is protected by service ground-fault protection.

II. Location

240.20 Ungrounded Conductors

An overcurrent device (fuse or circuit breaker) has to be connected in series with each ungrounded conductor. An overcurrent relay with a current transformer is allowed. Generally, a circuit breaker has to open all ungrounded conductors. Single-pole circuit breakers are permitted under certain conditions listed in this section. In a closed-loop power distribution system, listed devices can be used instead of fuses or circuit breakers.

240.21 Location in Circuit

This section lists the location of the overcurrent device in relation to the ungrounded circuit conductor.

(A) Branch-Circuit Conductors. If a branch-circuit tap meets the requirements of Section 210.19, then the overcurrent protection can be located in accordance with that Section.

(B) Feeder Taps

(1) Taps Not over 3 m (10 Feet) Long. They may be tapped to a feeder or transformer secondary without overcurrent protection when the tap is not longer than 3 m (10 ft.); the rating of the tap conductor is not less than the loads and devices it supplies and the overcurrent device at the end of the tap; the tap ends at the device it supplies; the tap conductors are enclosed in a raceway, and for field installations where the tap leaves the enclosure of the tap the rating of the overcurrent device on the line side of the tap is not more than 10 times the ampacity of the conductors.

(2) Taps Not over 7.5 m (25 Feet) Long. When the tap is not longer than 7.5 m (25 ft.) it does not require protection if the ampacity of the tap is at least one-third the rating of the feeder or overcurrent protection it was tapped from; ends in a single overcurrent device rated for the device; and is mechanically protected.

(3) Taps Supplying a Transformer [Primary Plus Secondary Not over 7.5 m (25 Feet) Long]. The rule is similar to the 7.5 m (25 ft.) tap (item c above). No overcurrent protection is required for a tap for a feeder supplying a transformer if the ampacity of the primary wires is at least one-third the rating of the overcurrent device protecting the feeder; when you multiply the rating of the conductors supplying the secondary by the ratio of the secondary to primary voltage it should be at least one-third the feeder protective overcurrent device rating; if both primary and secondary wires are not longer than 7.5 m (25 ft.) combined and both ratings (including transformer ratio for secondary) are not larger than one-third the rating of the circuit being tapped; the tap is mechanically protected; and the secondary conductors end in a single overcurrent device limiting the load current to that permitted by Section 310.15.

(4) Taps over 7.5 m (25 Feet) Long. No overcurrent protection is required for a tap located in manufacturing installations with bays over 11 m (35 ft.) in height if the horizontal length is not longer than 7.5 m (25 ft.) with the overall length not longer than 30 m (100 ft.). All the rules of the 7.5 m (25 ft.) tap rule (c) apply, and in addition, the tap conductors must be at least 6 AWG copper or 4 AWG aluminum; they can not go through floors, walls, or ceilings; the tap must be least 9 m (30 ft.) from the floor; the tap conductors have no splices and are continuous from one end to the other and only qualified persons service the system.

(5) Outside Taps of Unlimited Length

(C) Transformer Secondary Conductors. There is a complete discussion of transformer taps. Among the items discussed are protection by primary overcurrent device, transformer secondary conductors not over 3 m (10 ft.) long, secondary conductors not over 7.5 m (25 ft.) long, outside secondary conductors, and secondary conductors from a feeder tapped transformer.

(D) Service Conductors

(E) Busway Taps. Busways and their taps can be protected in accordance with Sections 364.10 through 364.14.

(F) Motor Circuit Taps. Motor branch circuit conductors can be protected as indicated in Sections 430.28 and 430.53.

(G) Conductors from Generator Terminals. They can be protected as indicated in Section 445.4 if sized as per 445.5.

240.22 Grounded Conductors

No overcurrent device should be placed in series with a grounded conductor unless it simultaneously opens all other conductors of the circuit or when required by Sections 430.36 and 430.37.

240.23 Change in Size of Grounded Conductor

240.24 Location in or on Premises

Overcurrent devices must be readily accessible with four exceptions. Each occupant must have access to the ones protecting the conductors for his or her occupancy. The overcurrent devices must not be exposed to physical damage and must not be located near easily ignitable material or in bathrooms of dwelling units or guest rooms of hotels and motels. Refer to the *NEC*® for certain exceptions.

III. Enclosures

240.30 General

240.32 Damp or Wet Locations

240.33 Vertical Position

IV. Disconnecting and Guarding

240.40 Disconnecting Means for Fuses

In general, the disconnection means must be provided on the supply side of the fuse. This does not hold true for service current-limiting devices as per Section 230.82 or single disconnecting devices for multiple sets of fuses as per Section 430.112 and Section 424.22(C).

240.41 Arcing or Suddenly Moving Parts

V. Plug Fuses, Fuseholders, and Adapters

240.50 General

These are not permitted in circuits where the voltage between conductors is more than 125 V except with a grounded neutral and not more than 150 V to ground. There are other requirements.

240.51 Edison-Base Fuses

240.52 Edison-Base Fuseholders

240.53 Type S Fuses

240.54 Type S Fuses, Adapters, and Fuseholders

VI. Cartridge Fuses and Fuseholders

240.60 General

These are permitted in circuits operating at 300 V between conductors if rated 300 V and for single-phase line to neutral circuits where this voltage is not larger than 300 V and the circuit is supplied from a three-phase, four-wire, solidly grounded neutral system. The holders must prevent interchangeability where a fuse can be put into a circuit of lower ampere rating or higher voltage rating. All fuses must be marked, indicating the current, voltage, and interrupting rating, if it is current limiting, and the manufacturer's name and trademark. The interrupting rating is not required when they are used for supplementary protection.

240.61 Classification

VII. Circuit Breakers

240.80 Method of Operation

Normally, they must be trip free and capable of manual operation in addition to other normal means, such as electrical or pneumatic operation.

240.81 Indicating

Must indicate on or off—on is up position if mounted vertically.

240.82 Nontamperable

240.83 Marking

They must all be properly marked. When properly used as a switch, they must be approved for switching duty and marked SWD or HID. There are requirements for voltage marking, interrupting ratings, and location.

240.85 Applications

240.86 Series Rating

VIII. Supervised Industrial Locations

240.90 General

240.92 Location in Circuit

IX. Overcurrent Protection over 600 Volts, Nominal

240.100 Feeders and Branch Circuits

240.101 Additional Requirements for Feeders

ARTICLE 250 GROUNDING

I. General

250.1 Scope

Generally gives requirements for grounding and bonding.

250.2 Definitions

250.3 Application of Other Articles

The following other articles have special grounding requirements:

TABLE 250.3 Additional Grounding Requirements

Conductor/Equipment	*Article*	*Section*
Agricultural Buildings		547.9 and 547.10
Audio Signal Processing, Amplification, and Reproduction Equipment		640.7

TABLE 250.3 Additional Grounding Requirements (continued)

Conductor/Equipment	*Article*	*Section*
Branch Circuits		210.5 210.6 406.3
Cablebus		370.9
Capacitors		460.10 460.27
Circuits and Equipment Operating at Less Than 50 Volts	720	
Class 1, Class 2, and Class 3 Circuits Remote Control, Signaling, and Power Limited Circuits		725.6
Closed-Loop and Programmed Power Distribution		780.3
Communications Circuits	800	
Community Antenna Television and Radio Distribution Systems		820.33 820.40 820.41
Conductors for General Wiring	310	
Cranes and Hoists	610	
Electrically Driven or Controlled Irrigation Machines		675.11(C) 675.12 675.13 675.14 675.15
Electric Signs and Outline Lighting	600	
Electrolytic Cells	668	
Elevators, Dumbwaiters, Escalators, Moving Walks, Wheel Chair Lifts, and Stairway Chair Lifts	620	
Fire Alarm Systems		760.6
Fixed Electric Heating Equipment for Pipelines and Vessels		427.29 427.48
Fixed Outdoor Electric De-icing and Snow-Melting Equipment		426.27

TABLE 250.3 Additional Grounding Requirements (continued)

Conductor/Equipment	*Article*	*Section*
Flexible Cords and Cables		400.22 400.23
Floating Buildings		553.8 553.10 553.11
Grounding-Type Receptacles, Adapters, Cord Connectors, and Attachment Plugs		406.9
Hazardous (Classified) Locations	500–517	
Health Care Facilities	517	
Induction and Dielectric Heating Equipment	665	
Industrial Machinery	670	
Information Technology Equipment		645.15
Intrinsically Safe Systems		504.50
Luminaires (lighting fixtures) and Lighting Equipment		410.17 410.18 410.20 410.21 410.105(B)
Luminaires (lighting fixtures), Lampholders, Lamps, and Receptacles	410	
Marinas and Boatyards		555.15
Mobile Homes and Mobile Home Parks	550	
Motion Picture and Television Studios and Similar Locations		530.20 530.66
Motors, Motor Circuits, and Controllers	430	
Outlet, Device, Pull and Junction Boxes, Conduit Bodies, and Fittings		314.4 314.25
Over 600 Volts, Underground Wiring Methods		350.50(B)
Panelboards		408.20
Pipe Organs	650	
Radio and Television Equipment	810	

TABLE 250.3 Additional Grounding Requirements (continued)

Conductor/Equipment	*Article*	*Section*
Receptacles and Cord Connectors		406.3
Recreational Vehicles and Recreational Vehicle Parks	551	
Services	230	
Solar Photovoltaic Systems		690.41 690.42 690.43 690.45 690.47
Swimming Pools, Fountains, and Similar Installations	680	
Switchboards and Panelboards		408.3(D)
Switches		404.12
Theaters, Audience Areas of Motion Picture and Television Studios, and Similar Locations		520.81
Transformers and Transformer Vaults		450.10
Use of Identification of Grounded Conductors	200	
X-Ray Equipment	660	517.78

250.4 General Requirements for Grounding and Bonding

(A) Grounded Systems

(1) Electrical System Grounding. If the system is to be grounded, it has to be connected to earth so that it will limit voltages due to lightning, unintentional contact with higher voltages, and line surges. It must also stabilize line voltages during normal operation.

(2) Grounding of Electrical Equipment. Any non-current-carrying conductive material enclosing electrical conductors or equipment, or part of a piece of equipment must be connected to earth.

(3) Bonding of Electrical Equipment. All non-current carrying conductive material which encloses conductors or equipment must be connected together and to the electrical supply source.

(4) Bonding of Electrically Conductive Materials and Other Equipment. Any electrically conductive material that may become energized has to be bonded together and to the supply system at the source.

(5) Effective Ground Fault Current Path. The fault current path has to carry the maximum fault current, have low impedance and be continuous and permanent in nature.

The earth is not to be used as the only equipment-grounding conductor or fault current path.

(B) Ungrounded Systems

(1) Grounding of Electrical Equipment. Non-current carrying conductive materials that enclose conductors or equipment must be connected to earth so that it will limit voltages due to lightning, unintentional contact and will limit the voltage to ground.

(2) Bonding of Electrical Equipment. All materials that enclose conductors or equipment must be connected together and to the supply system grounded equipment so that the fault current path carries maximum fault current, has low impedance and is continuous and permanent in nature. It must be capable of carrying the maximum fault current that can occur.

(3) Bonding of Electrically Conductive Materials and Other Equipment. Any electrically conductive material that can become energized has to be bonded together and to the supply system grounding equipment so that the fault current path carries maximum fault current, has low impedance, and is continuous and permanent in nature. It must be capable of carrying the maximum fault current that can occur.

(4) Path for Fault Current. The fault current path has to carry the maximum fault current path and have low impedance from any place on the system so that the overcurrent devices will operate on a second fault.

There are two fine print notes.

250.6 Objectionable Current over Grounding Conductors

The entire electrical system is to be installed to prevent an objectionable flow of current over the grounding system. There is a discussion of alterations to stop objectionable current, limitations to permissible alterations, isolation of objectionable direct-current ground currents, and the fact that temporary currents are not classified as objectionable currents.

250.8 Connection of Grounding and Bonding Equipment

250.10 Protection of Ground Clamps and Fittings

250.12 Clean Surfaces

II. Circuit and System Grounding

250.20 Alternating-Current Circuits and Systems to Be Grounded

The following alternating-current circuits and systems must be grounded. Others may be grounded if desired.

(A) Alternating-current circuits of less than 50 V must be grounded if supplied by transformer systems of more than 150 V to ground or by ungrounded transformer systems. All outdoor overhead circuits must be grounded.

(B) Alternating-current circuits between 50 and 1000 V of the following types must be grounded:

1. Systems of up to 150 V to ground from an ungrounded conductor
2. Systems that are three-phase, four-wire with a neutral circuit conductor
3. Three-phase, four-wire delta systems with the midpoint of one phase used as a circuit conductor

(C) Alternating-Current Systems 1 kV and Over

(D) Separately Derived Systems. If the source is separately derived, and covered in (A) or (B), it must be grounded as required by Section 250.30.

250.21 Alternating Current Systems of 50 Volts to 1000 Volts Not Required to Be Grounded

There are a number of systems that are permitted to be grounded but are not required to be grounded. They are: electric systems used to supply industrial electric furnaces for melting, etc.; separately derived systems used only to supply rectifiers for adjustable speed industrial drives; high-impedance grounded neutral systems noted in Section 250.36, other systems in accordance with 250.20(B) and separately derived systems supplied by transformers less than 1000 volts provided that special conditions are met. There are four of these special conditions.

250.22 Circuits Not to Be Grounded

Those specified in Article 517, electric cranes used in areas of Class III over combustible fibers as indicated in Section 503.13, Electrolytic Cells noted in Article 668 and lighting systems secondary circuits as noted in Section 411.5(A).

250.24 Grounding Service-Supplied Alternating-Current Systems

As required, each service must have a grounding electrode conductor connected to the grounded service conductor. The reader should be familiar with the requirements of this article in the *NEC®*. There is a discussion of system grounding connections in general and specifically from outdoor transformers, dual-fed services, using the main bonding jumper as wire or busbar, and loadside grounding connections. There is also a discussion of systems where the grounded conductor is brought to the service equipment, ungrounded system grounding connections, as well as the grounding electrode conductor.

250.26 Conductor to Be Grounded—Alternating-Current Systems

One conductor must be grounded in a two-wire, single-phase system. The neutral conductor is grounded in a three-wire, single-phase system and a multiphase system where one phase is used as a three-wire, single-phase system. The common conductor is grounded in a multiphase system with one wire common to all phases. A phase conductor is grounded in a multiphase system where one phase is grounded and the neutral conductor is grounded in a multiphase system where one phase is used as a single phase 3 wire system.

250.28 Main Bonding Jumper

250.30 Grounding Separately Derived Alternating-Current Systems

250.32 Two or More Buildings or Structures Supplied from a Common Service

Each building must have a grounding electrode connected to the building grounding means. There are methods noted for grounded systems, ungrounded systems, where the disconnecting means is located in a separate building or structure on the same premises, and the size of the grounding conductor. The size of grounding conductor to the grounding electrode is given in Table 25.122. The reader should refer to this section in the *NEC*®.

250.34 Portable and Vehicle-Mounted Generators

250.36 High-Impedance Grounded Neutral Systems

III. Grounding Electrode System and Grounding Electrode Conductor

250.50 Grounding Electrode System

If it is available each item delineated in 250.52(A)(1) through (A)(6) has to be bonded together to form the grounding electrode system. If none is available then at least one of the electrodes in 252(A)(4) through (A)(7) must be used.

250.52 Grounding Electrodes

(A) Electrodes Permitted for Grounding

(1) Metal Underground Water Pipe. It must be at least in contact with the earth for 3 m (10) or more feet. It must be electrically continuous to the point of connection to the grounding electrode system and bonding conductors. Interior water piping can not be used if it is more than 1.52 m (5 feet) from the point of entrance to the building. There is an exception for industrial and commercial buildings under certain conditions.

(2) Metal Frame of the Building or Structure. Can be used if effectively grounded.

(3) Concrete-Encased Electrode. The electrode must be at least 6 m (20 ft.) of 4 AWG bare copper conductor or 6 m (20 ft.) of steel or zinc galvanized reinforcing bars or roads at least 12.7 mm (½ in.) in diameter. It must be encased in at least 50 mm (2 in.) of concrete within or near the bottom of the foundation.

(4) **Ground Ring**

(5) **Rod and Pipe Electrodes**

(6) **Plate Electrodes**

(7) **Other Local Metal Underground Systems or Structures**

(B) Electrodes Not Permitted for Grounding. Metal underground gas piping and aluminum electrodes are not permitted.

250.53 Grounding Electrode System Installation

(A) **Rod, Pipe, and Plate Electrodes**

(B) **Electrode Spacing**

(C) **Bonding Jumper**

(D) **Metal Underground Water Pipe**

(E) **Supplemental Electrode Bonding Site**

(F) **Ground Ring**

(G) **Rod and Pipe Electrodes**

(H) **Plate Electrode**

250.54 Supplementary Grounding Electrodes

They are permitted to be connected to the equipment grounding conductors specified in Section 250.118. The earth can not be used as the only equipment grounding conductor. They are not required to comply with bonding requirements of 250.50 or resistance requirements of 250.56.

250.56 Resistance of Rod, Pipe, and Plate Electrodes

250.58 Common Grounding Electrode

Where an ac system is grounded, the system and equipment must be grounded to the same electrode or electrodes bonded together.

250.60 Use of Air Terminal

You cannot use air terminals conductors and driven pipes, rods, or plate electrodes used for grounding air terminals in lieu of grounding electrodes as noted in Section 250.50. You can bond grounding electrodes of different systems.

250.62 Grounding Electrode Conductor Material

The material is to be solid or stranded copper, aluminum, or copper-clad aluminum. It can be bare, stranded, covered, or insulated. It must be corrosion resistant.

250.64 Grounding Electrode Conductor Installation

(A) Aluminum or Copper-Clad Aluminum Conductors. Where in direct contact with masonry or the earth where there are corrosive conditions, insulated or bare aluminum or copper-clad aluminum grounding conductors are not permitted. When used outside, these types of conductors cannot be used within 450 mm (18 in.) of the earth. The installation must be as noted below.

(B) Securing and Protection from Physical Damage. If it is 4 AWG copper or aluminum or larger, it must be protected if it is subject to physical damage that is severe. 6 AWG wires, if fastened to the building and free from physical damage, do not need protection; if not, they must be protected in the same manner as wires smaller than 6 AWG. Wires smaller than 6 AWG must be enclosed in either rigid metal conduit, intermediate metal conduit, rigid nonmetallic conduit, electrical metallic tubing, or cable armor. In all cases above, the wire or enclosure must be fastened to the surface it is running on. Insulated or bare aluminum or copper-clad aluminum cannot be used in corrosive areas.

(C) Continuous. The grounding electrode cannot have a splice or joint. A splice will be permitted only if an irreversible compression-type connector is used that is listed for the use or by exothermic welding process.

(D) Grounding Electrode Conductor Taps. Taps to the grounding electrode conductor are permitted where a service has more than one enclosure as noted in Section 230.40, Exception No. 2. The taps must go inside of the enclosure. The grounding electrode conductor has to be sized in accordance with Section 250.66. The tap conductors can be sized for the largest conductor. The tap conductor has to be connected to the grounding electrode conductor in such a way as to maintain the grounding electrode conductor without a splice.

(E) Enclosures for Grounding Electrode Conductors. A complete electrical path must be ensured for metal enclosures of grounding electrode conductors. If a raceway is used as protection for a grounding conductor, compliance is required with the appropriate article concerning the raceway.

(F) To Electrodes. A grounding conductor can be run to any convenient grounding electrode in the grounding electrode system or to one or more grounding electrodes. It must be sized for the largest grounding electrode required by the electrodes connected to it.

250.66 Size of Alternating-Current Grounding Electrode Conductor

These minimum sizes are given in Table 250.66. There are three special cases.

250.68 Grounding Electrode Conductor Connection to Grounding Electrodes

It must be accessible and provide an effective and permanent ground.

250.70 Methods of Grounding Conductor Connections to Electrodes

IV. Enclosure, Raceway, and Service Cable Grounding

250.80 Service Raceways and Enclosures

These metal enclosures and raceways must be grounded. There is an exception for a metal elbow installed in an underground installation under certain conditions.

250.84 Underground Service Cable or Conduit

The metal sheath which is metallically connected to the underground systems of an underground cable system or its metal conduit does not have to be grounded at the building, assuming that it is a continuous system.

250.86 Other Conductor Enclosures and Raceways

These metal enclosures and raceways must be grounded except as permitted in Section 250.112(I). Note three exceptions: for open wire, knob and tubing, and nonmetallic-sheathed cable; enclosures used to physically protect cable assemblies; and metal elbows under certain conditions.

V. Bonding

250.90 General

Bonding must be provided to make sure there is proper electrical continuity in the grounding system.

250.92 Services

(A) Bonding at Services

All service equipment must be bonded, items are listed. A means must be provided for intersystem bonding.

TABLE 250.66 Grounding Electrode Conductor for Alternating-Current Systems

Size of Largest Ungrounded Service-Entrance Conductor or Equivalent Area for Parallel Conductors[1] (AWG/kcmil)		*Size of Grounding Electrode Conductor (AWG/kcmil)*	
Copper	*Aluminum or Copper-Clad Aluminum*	*Copper*	*Aluminum or Copper-Clad Aluminum*[2]
2 or smaller	1/0 or smaller	8	6
1 or 1/0	2/0 or 3/0	6	4
2/0 or 3/0	4/0 or 250	4	2
Over 3/0 through 350	Over 250 through 500	2	1/0
Over 350 through 600	Over 500 through 900	1/0	3/0
Over 600 through 1100	Over 900 through 1750	2/0	4/0
Over 1100	Over 1750	3/0	250

Where multiple sets of service-entrance conductors are used as permitted in Section 230.40, Exception No. 2, the equivalent size of the largest service-entrance conductor shall be determined by the largest sum of the areas of the corresponding conductors of each set. Where there are no service-entrance conductors, the grounding electrode conductor size shall be determined by the required for the load to be served.

[1]This table applies to the derived conductors of separately derived ac systems.

[2]See installation restrictions in Section 250.64(A).

(B) Method of Bonding at the Service

There are four approved methods:

(1) For the grounded service conductor use a method noted in Section 250.8

(2) Threaded couplings or threaded bosses made up wrench tight

(3) Threadless couplings and connectors for where made tight for metal raceways and metal-clad cables

(4) Other approved devices

250.94 Bonding for Other Systems

250.96 Bonding Other Enclosures

All items that will be used as grounding conductors must be bonded. There is a provision for isolated grounding circuits.

250.97 Bonding for over 250 Volts

250.98 Bonding Loosely Jointed Metal Raceways

250.100 Bonding in Hazardous (Classified) Locations

All items in a hazardous (classified) location must be bonded as required by Section 250.94, regardless of the voltage.

250.102 Equipment Bonding Jumpers

250.104 Bonding of Piping Systems and Exposed Structural Steel

250.106 Lightning Protection Systems

These ground terminals have to be bonded to the building ground system.

VI. Equipment Grounding and Equipment Grounding Conductors

250.110 Equipment Fastened in Place or Connected by Permanent Wiring Methods (Fixed)

Metal parts of fixed equipment that are likely to become energized must be grounded if located in a wet or damp location and not isolated, if they can be touched by people and are 2.5 m (8 ft.) or less vertically and 1.5 m (5 ft.) or less horizontally to ground or a grounded object, if in contact with metal, if in a hazardous location, if metal-clad type of wiring is used under certain conditions, and if the voltage is greater than 150 V to ground. There are three exceptions.

250.112 Fastened in Place or Connected by Permanent Wiring Methods (Fixed)—Specific

All exposed metal parts that do not conduct current of the following kinds of equipment and enclosures must be grounded: switchboard frames and structures except two-wire dc where insulated from ground: pipe organs; motor frames; enclosures for motor controllers; elevators and cranes; all electric equipment in commercial garages, theaters, and motion picture studios except pendant lampholders of 150 V or less: electric signs, outline lighting and associated equipment as noted in Article 600; motion picture equipment; power-limited remote-control, signaling, and fire-alarm signaling equipment supplied by Class 1, 2, and 3 circuits where required; luminaires (lighting fixtures) as per Article 410; skid-mounted equipment, motor-operated water pumps, and metal well casings.

250.114 Equipment Connected by Cord and Plug

There are areas where exposed metal parts (noncurrent-carrying) must be grounded even though connected by cord and plug. This is true of hazardous locations and circuits over 150 V to ground; two exceptions are motors when guarded and electrically heated appliances with special exemptions and insulation to ground. In a residence the following items are required to have equipment grounding: refrigerators, freezers, and air conditioners; clothes washing and clothes drying equipment, dishwashing machines, kitchen waste disposers, information technology equipment, sump pumps, and electrical aquarium equipment; hand-held motor-operated tools, stationary and fixed motor-operated tools, light industrial motor-operated tools; motor-operated hedge clippers, lawn mowers, snow blowers, wet scrubbers, etc.; and portable handlamps. In other than residences, the list is the same as residences, with the addition of cord- and plug-connected appliances in damp or wet locations and tools used in wet or conductive locations. There is an exception for tools and portable handlamps in wet or conductive locations if supplied through an isolating transformer with a secondary ungrounded and not over 50 V.

250.116 Nonelectric Equipment

The metal parts of this equipment must be grounded: cranes and hoists; elevator cars of nonelectric type; and electric elevators.

250.118 Types of Equipment Grounding Conductors

The following may be used for equipment grounding conductors.

1. Copper or other corrosion-resistant conductor. It can be solid or stranded, and bare, covered, or insulated.
2. Rigid metal conduit.
3. Intermediate metal conduit.
4. Electrical metallic tubing.
5. Flexible metal conduit when the fittings and conduit are listed for grounding.
6. Listed flexible metal conduit that is not listed for grounding but meets additional requirements. These are that the terminations are listed for grounding; the conductors in the conduit are protected by devices that are rated 20 A or less; the total length of flexible metal conduit, flexible metallic tubing, and liquidtight flexible metal conduit in the same ground return path is not more than 1.8 m (6 ft.); and the conduit is not installed because it is flexible.
7. Listed liquidtight flexible metal conduit meeting certain requirements.
8. Flexible metal tubing under certain conditions.
9. Armor of Type AC cable.
10. The copper sheath of MI metal sheathed cable.
11. Type MC cable under certain conditions.
12. Cable trays as noted in Sections 318.3(c) and 318.7.
13. Cablebus framework and noted in Section 365.2(A).
14. Any other metal raceway and auxiliary gutter that is continuous and listed for grounding.

250.119 Identification of Equipment Grounding Conductors

The equipment grounding conductor can either be bare, insulated, or covered unless there is a specific required by the code. There are identification requirements for conductors. Special requirements are listed for conductors larger than 6 AWG, multiconductor cable, and flexible cord.

250.120 Equipment Grounding Conductor Installation

Where it is a raceway, cable tray, cable armor, cable bus, or cable sheath, or a wire in a raceway it must be installed in accordance with the *NEC*®. Aluminum and copper-clad aluminum can be used under certain circumstances. If the equipment grounding conductor is smaller than 6 AWG, it has to be protected from physical damage except under certain conditions where it is not subject to physical damage.

250.122 Size of Equipment Grounding Conductors

These minimum sizes are given in Table 250.122. When circuit conductors are run in parallel, the equipment grounding conductor must be run in parallel and sized according to the overcurrent device rating shown in the table. The conductor never has to be larger than the overcurrent device protecting the circuit conductor to the equipment. There are other requirements and exceptions, including where conductors are adjusted in size to compensate for voltage drop.

250.124 Equipment Grounding Conductor Continuity

250.126 Identification of Wiring Device Terminals

VII. Methods of Equipment Grounding

250.130 Equipment Grounding Conductor Connections

250.132 Short Sections of Raceway

250.134 Equipment Fastened in Place or Connected by Permanent Wiring Methods (Fixed)—Grounding

250.136 Equipment Considered Effectively Grounded

(A) This is true when it is electrically connected to a supporting metal rack or structure that is properly grounded. The building's structural steel is not considered an effective grounding method.

(B) Metal car frames

250.138 Cord- and Plug-Connected Equipment

250.140 Frames for Ranges and Clothes Dryers

250.142 Use of Grounded Circuit Conductor for Grounding Equipment

(A) Generally, a grounded circuit conductor can be used to ground equipment on the supply side of the service disconnect, supply side of main disconnects for a separate building [Section 250.32(B)], and in a separate, derived system the supply side of the disconnecting or overcurrent devices.

TABLE 250.122 Minimum Size Equipment Grounding Conductors for Grounding Raceway and Equipment

Rating or Setting of Automatic Overcurrent Device in Circuit Ahead of Equipment, Conduit, etc., Not Exceeding (Amperes)	*Size (AWG or kcmil)*	
	Copper	*Aluminum or Copper-Clad Aluminum*[1]
15	14	12
20	12	10
30	10	8
40	10	8
60	10	8
100	8	6
200	6	4
300	4	2
400	3	1
500	2	1/0
600	1	2/0
800	1/0	3/0
1000	2/0	4/0
1200	3/0	250
1600	4/0	350
2000	250	400
2500	350	600
3000	400	600
4000	500	800
5000	700	1200
6000	800	1200

[1]See installation restrictions in Section 250.120.

Note: Where necessary comply with Section 250.4(A)(5) or 250.4(B)(4), the equipment grounding conductor shall be sized larger than this table.

(B) Generally, a grounded circuit conductor cannot be used to ground equipment on the load side of the service disconnect. There are three exceptions to this:

1. As indicated in Section 250.140 for ranges, wall-mounted ovens, etc.
2. Grounding meter enclosures if no service ground-fault protection is used, the meter is near the service switch, and the grounded circuit conductor is not smaller than noted in Table 250.122 for equipment grounding conductors.
3. Dc systems in accordance with Section 250.22, Exception.
4. Ground electrode-type boilers over 600 V in accordance with 490.72(E)(1) and 490.74.

250.144 Multiple Circuit Connections

250.146 Connecting Receptacle Grounding Terminal to Box

Generally, a bonding jumper must be used unless grounded as follows:

A. Surface Mounted Box. If the box is surface mounted and the yoke makes good electrical contact with the box. There are restrictions on cover-mounted receptacles.

B. Contact Devices or Yokes. Where the receptacle is listed as such with special yokes and screws.

C. Floor Boxes. Floor boxes so listed.

D. Isolated Receptacles. Where there is a requirement for the reduction of electrical noise, an insulated grounding conductor can be used. It is run with the circuit conductors. It can pass through panelboards without connection to the grounding terminal. The grounding terminal in the receptacle can be insulated from the mounting.

250.148 Continuity and Attachment of Equipment Grounding Conductors to Boxes

VIII. Direct Current Systems

250.160 General

250.162 Direct-Current Systems and Systems to Be Grounded

250.164 Point of Connection for Direct-Current Systems

250.166 Size of Direct-Current Grounding Electrode Conductor

It cannot be smaller than the neutral conductor for a three-wire balancer set or balancer winding. It cannot be smaller than the largest conductor if other than as stated in the preceding sentence. It can never be smaller than 8 AWG copper or 6 AWG aluminum. There are requirements for connections to make electrodes, concrete-encased electrodes, and ground rings.

250.168 Direct-Current Bonding Jumper

250.169 Underground Direct-Current Separately Derived Systems

IX. Instruments, Meters, and Relays

250.170 Instrument Transformer Circuits

250.172 Instrument Transformer Cases

250.174 Cases of Instruments, Meters, and Relays Operating at Less Than 1000 Volts

250.176 Cases of Instruments, Meters and Relays Operating at 1 kV and Over

250.178 Instrument Grounding Conductor

X. Grounding of Systems and Circuits of 1 kV and over (High Voltage)

250.180 General

If these systems are grounded, they must follow all previous sections as well as this section.

250.182 Derived Neutral Systems

250.184 Solidly Grounded Neutral Systems

250.186 Impedance Grounded Neutral Systems

250.188 Grounding of Systems Supplying Portable or Mobile Equipment

250.190 Grounding of Equipment

ARTICLE 280 SURGE ARRESTERS

I. General

280.1 Scope

280.2 Definition

280.3 Number Required

280.4 Surge-Arrester Selection

II. Installation

280.11 Location

280.12 Routing of Surge-Arrester Connections

III. Connecting Surge Arresters

280.21 Installed at Services of Less Than 1000 Volts

280.22 Installed on the Load Side of Services of Less Than 1000 Volts

280.23 Circuits of 1 kV and over—Surge-Arrester Conductors

280.24 Circuits of 1 kV and over—Interconnections

280.25 Grounding

ARTICLE 285
Transient Voltage Surge Suppressors (TVSS)

I. General

285.1 Scope

285.2 Definition

285.3 Uses Not Permitted

285.4 Number Required

285.5 Listing

285.6 Short Circuit Current Rating

II. Installation

285.11 Location

280.12 Routing of Connections

III. Connecting Transient Voltage Surge Suppressors

285.21 Connection

280.25 Grounding

3

WIRING METHODS AND MATERIALS

ARTICLE 300
WIRING METHODS

I. General Requirements

300.1 Scope

This article applies to all wiring installations except as noted in other articles. It does not apply to integral parts of equipment (e.g., motors, motor control centers, etc.) Metric designators and trade sizes for conduit, tubing, etc. are shown in Table 300.1(C).

300.2 Limitations

This chapter applies to voltages up to and including 600 V when not otherwise noted and over 600 V only when specifically mentioned in the *National Electrical Code*®.

300.3 Conductors

(A) Single Conductors

(B) Conductors of the Same Circuit. All conductors of the same circuit including the grounded conductor and equip-

TABLE 300.1(C) Metric Designator and Trade Sizes

Metric Designator	*Trade Size*
12	⅜
16	½
21	¾
27	1
35	1¼
41	1½
53	2
63	2½
78	3
91	3½
103	4
129	5
155	6

Note: The metric designators and trade sizes are for identification purposes only and are not actual dimensions.

ment grounding conductors must be installed in the same raceway, cable, etc. There are exceptions.

(C) Conductors of Different Systems

(1) 600 Volts, Nominal, or Less. Conductors of 600 V or less can occupy the same wiring enclosure if all conductors have insulation rated for the maximum voltage used. Refer to Section 690.4(B) for solar photovoltaic systems and to Section 725.55(A) for Class 2 and Class 3 conductors.

(2) Over 600 Volts, Nominal. Conductors of over 600 V cannot be in the same wiring enclosure with those rated 600 V or less except for secondary wiring to electric discharge lamps, primary leads of electric discharge lamp ballasts, and excitation, control, relay, ammeter conductors, motor switch gear and control assemblies, and in manholes under certain conditions.

Reference should be made to this section in the *NEC*® for a complete explanation.

300.4 Protection against Physical Damage

300.5 Underground Installations

(A) Minimum Cover Requirements. The cover requirements for direct buried cable or conduit are shown in Table 300.5. The reader should refer to it.

(B) Grounding

(C) Underground Cables under Buildings

(D) Protection from Damage

(E) Splices and Taps

(F) Backfill

(G) Raceway Seals

(H) Bushing

(I) Conductors of the Same Circuit

(J) Ground Movement

(K) Directional Boring

300.6 Protection against Corrosion

Generally, all material used must be suitable for use in the installed environment.

(A) General. Ferrous material must be protected both inside and out by a coating of corrosion-resistant material (e.g., zinc, cadmium, or enamel). Enamel cannot be used outdoors or in wet locations. A specific exception is made for threads at joints.

(B) In Concrete or in Direct Contact with the Earth. If the material is made for this purpose, it can be used.

(C) Indoor Wet Locations. The cable must be installed with at least 6 mm (¼ in.) of air space from the wall or supporting surface when installed exposed. There is an exception.

300.7 Raceways Exposed to Different Temperatures

300.8 Installations of Conductors with Other Systems

300.10 Electrical Continuity of Metal Raceways and Enclosures

There must be electrical continuity between all metal race-ways, cable armor, and metal enclosures. There are two exceptions.

300.11 Securing and Supporting

300.12 Mechanical Continuity—Raceways and Cables

300.13 Mechanical and Electrical Continuity—Conductors

Splices or taps are not permitted within a raceway. Exceptions to this are noted in Sections 300.15; 368.8(A); 376.56; 378.56;

384.56; 386.56; 388.56; or 390.6. In a multiwire branch circuit the continuity cannot depend on device connections that can be removed.

300.14 Length of Free Conductors at Outlets, Junctions, and Switch Points

At least 150 mm (6 in.) of conductors must be left at every outlet, junction, or switch point for splices or connections. When the opening to the point of connection is less than 200 mm (8 in.), the conductors must extend at least 75 mm (3 in.).

300.15 Boxes, Conduit Bodies or Fittings—Where Required

For concealed knob and tube wiring a box must be installed at each switch point and outlet. Use fittings and connectors only for the purpose that they are listed and designed for. If the wiring method is conduit, tubing, Type AC cable, Type MC cable, Type MI cable, non-metallic sheathed cable or other cables, then a conduit body or box must be installed at each conductor splice point, switch point, outlet point, termination point, pull point and junction point. The conduit body or box must meet the requirements of Article 314. This requirement for the use of a conduit body or box is in force unless not required by (A) through (M) following.

(A) Wiring Methods With Interior Access. If the outlet method has a removable cover that is accessible after installation.

(B) Equipment. A wiring compartment or junction box that is part of approved equipment is allowed instead of a box.

(C) Protection. If the cables enter or leave conduit or tubing that is used for support or protection against physical damage. There are requirements for the use of fittings.

(D) Type MI Cable. If accessible fittings are used for straight through splices.

(E) Integral Enclosure. A wiring device with an integral enclosure that has brackets that fasten it to wall or ceilings for use with nonmetallic sheathed cable is permitted.

(F) Fitting. A fitting can be used instead of a conduit body or box as long as there are splices or terminations in the fitting. The fitting has to be accessible after the installation.

(G) Direct-Buried Conductors. As permitted in 300.5(E).

(H) Insulated Devices. As permitted in 334.40(B) where the insulated device is supplied by nonmetallic-sheathed cable.

(I) Enclosures. Where a splice, switch terminal, or pull point is in a cabinet or cut out box in an enclosure for a switch or overcurrent device as noted in 312.8 in a motor controller as noted in 430.10(A) or in a motor center.

(J) Luminaires (Fixtures). If the fixture is used as a raceway as permitted in 410.31 and 410.32.

(K) Embedded. Where conductors are embedded as permitted in 424.40, 424.41(D), 426.22(B), 426.24(A), and 427.19(A) they are not required for splices.

(L) Manholes. If only qualified persons have access in manholes except when connecting to electrical equipment and compliance with Part IV of Article 314.

(M) Closed Loop. If a device is identified and listed for installation without a box and is used with a closed loop power distribution system.

300.16 Raceway or Cable to Open or Concealed Wiring

300.17 Number and Size of Conductors in Raceway

In general, the number and size of conductors in a raceway must allow the dissipation of heat, easy installation, and withdrawal.

300.18 Raceway Installations

Raceways other than exposed raceways or busways must be installed complete between outlet, junction, or splicing points before the conductors are installed. Three exceptions are: (1) for prewired assemblies, (2) when necessary for the installation of utilization equipment, and (3) exposed raceways with removable covers. There is a restriction on the use of welding.

300.19 Supporting Conductors in Vertical Raceways

Conductors must be supported at the top of vertical risers with additional support as noted in Table 300.19(A) of the *NEC*® when the rise is greater than the values in that table.

300.20 Induced Currents in Metal Enclosures or Metal Raceways

The effects of induced currents in metal enclosures must be minimized. This is done by grouping all phases, neutral, and grounding conductors in one raceway. If individual conductors are used, a slot can be cut between the holes made in the metallic enclosure through which the individual conductor passes. There are some exceptions.

300.21 Spread of Fire or Products of Combustion

300.22 Wiring in Ducts, Plenums, and Other Air-Handling Spaces

(A) Ducts for Dust, Loose Stock, or Vapor Removal. Wiring of any type is prohibited in these types of installations. This includes ventilation of commercial-type cooking equipment.

(B) Ducts or Plenums Used for Environmental Air. The following can be used when the duct or plenum was specifically made for this purpose: Type MI, Type MC cable with the specified sheath, EMT flexible metallic tubing, intermediate metal conduit, or rigid metal conduit. When necessary to connect physically adjustable items, regular or liquidtight flexible metal conduit can be used, but only in lengths not longer than 1.2 m (4 ft.). Only devices that act on or sense the contained air, such as thermostats, are permitted. If lighting fixtures are installed for maintenance or repair, they must be of the enclosed gasketed type. This paragraph applies primarily to fabricated ducts and plenums.

(C) Other Space Used for Environmental Air. This paragraph applies to spaces other than those in Sections 300.22(A) and (B) for environmental air handling. Habitable rooms or areas of buildings which are not primarily used for air handling are not included. Only totally enclosed nonventilated insulated busways that do not have a provision for plug-in connections and Type MI cable, Type MC cable without an overall nonmetallic covering, as well as Type AC cable or other factory-assembled multiconductor control or power cable made to be installed in these areas can be installed in this space. Other types of cables and conductors can be used but only when installed in EMT, flexible metallic tubing, intermediate metal conduit, rigid and flexible metal conduit, where accessible surface metal raceway or metal wireway with metal covers or solid bottom metal cable tray with solid metal covers. Also permitted is electric equipment with a metal enclosure or a listed nonmetallic enclosure with low smoke- and fire-resistive qualities suitable for the ambient temperature unless prohibited in another place in the Code.

(D) Information Technology Equipment. Refer to Article 645.

300.23 Panels Designed to Allow Access

II. Requirements for Over 600 Volts, Nominal

300.31 Covers Required

300.32 Conductors of Different Systems

See Section 300.3(C)(2).

300.34 Conductor Bending Radius

300.35 Protection against Induction Heating

300.37 Aboveground Wiring Methods

300.39 Braid-Covered Insulated Conductors-Open Installation

300.40 Insulation Shielding

300.42 Moisture or Mechanical Protection for Metal Sheathed Cables

300.50 Underground Installations

ARTICLE 310 CONDUCTORS FOR GENERAL WIRING

310.1 Scope

Article 310 is to be used for general requirements for wiring, but not in areas where it is part of an integral device such as a motor, motor controller, or where covered in another part of the *NEC*®.

310.2 Conductors

Conductors must be insulated and made of copper, copper-clad aluminum, or aluminum unless otherwise specified in the *NEC*®.

310.3 Stranded Conductors

Any conductor installed in a raceway must be stranded if it is 8 AWG or larger. This is not required if it is permitted or required in another part of the *NEC*®.

310.4 Conductors in Parallel

Conductors 1/0 AWG or larger can be run in parallel if they are the same length, the same size, and of the same material with the same insulation and terminated the same way. If run separately, the raceways must be the same. There are four exceptions to this:

1. Section 620.12(A)(1), Exception.
2. If smaller than 1/0 AWG, used for control power to indicating instruments, relays, and similar devices, and in the same raceway, each individual conductor can carry the entire load and the overcurrent rating is not larger than the ampacity of any single conductor.

3. Sizes smaller than 1/0 AWG are allowed for frequencies of 360 Hz and higher provided that Exception 2 is met.

4. With engineering supervision 2 AWG or larger grounded neutral conductors can be run parallel in existing installations.

310.5 Minimum Size of Conductors

Minimum sizes are shown in Table 310.5. For voltages up to 2000, it is 14 AWG for copper and 12 AWG for aluminum or copper-clad aluminum. The following exceptions apply:

1. Refer to Section 400.12 for flexible cords.
2. Fixture wire permitted by 402.6.
3. Refer to Section 430.22(F) for one horsepower or less motors.
4. Refer to Section 610.14 for cranes and hoists.
5. Refer to Section 620.12 for elevator control and signaling circuits.
6. Refer to Sections 725.27(A) and 725.51 for Class 1, 2, and 3 circuits.
7. Refer to Sections 760.27(A), 760.51, and 760.71(B) for fire alarm signaling circuits.
8. Refer to Section 430.72 for motor control circuits.
9. Refer to Section 727.6 for control and instrumentation circuits.
10. Refer to Sections 600.31(B) and 600.32(B) for electric signs and outline lighting.

310.6 Shielding

310.7 Direct Burial Conductors

310.8 Locations

In dry locations any type of insulated conductor or cable noted in the Code can be used. Use only lead-covered conductors; Type RHW, TW, THW, MTW, RHW-2, THW-2, THHW, THHW-2, THWN, THWN-2, XHHW, XHHW-2, ZW, or a type listed as such for a wet location, or moisture impervious metal-sheathed conductors can be used in wet locations. For dry and damp locations use Types FEP, FEPB, MTW, PFA, RHW, RHH, RHW-2, SA, XHH, XHHW, XHHW-2, THHN, THW-2, THW, THHW, THHW-2, THWN, THWN-2, TW, Z, or ZW. There is a note for locations exposed to direct sunlight.

310.9 Corrosive Conditions

310.10 Temperature Limitations of Conductors

310.11 Marking

All conductors and cables must have marked on them the maximum rated voltage, the proper letter type, the manufacturer's name or trademark, the AWG size or circular mil area, and cable assemblies where the neutral conductor is smaller than the ungrounded conductor. The size markings should be on the surface and spaced not more than 610 mm (24 in.) for eight types of cables. Other markings may be spaced 1.0 m (40 in.) apart. Certain cables can use marker tape inside the cable. Refer to this section in the *NEC*® for specific types where marker tape is permitted, which cables can have the size located elsewhere, and suffix designations for number of conductors. Special characteristics can be surface-marked on conductors listed in Chapter 3.

310.12 Conductor Identification

(A) Grounded Conductors. Insulated grounded conductors must be identified as noted in Section 200.6.

(B) Equipment Grounding Conductors. Equipment grounding conductors must be as noted in Section 250.119.

(C) Ungrounded Conductors. These must be distinguished from grounded and grounding conductors.

310.13 Conductor Constructions and Applications

Insulated conductors must be in accordance with Tables 310.13 and 310.61 through 310.64.

310.14 Aluminum Conductor Material

310.15 Ampacities for Conductors Rated 0-2000 Volts

(A) General

(1) Tables or Engineering Supervision. Ampacities can be determined by tables or under engineering supervision as noted in 310.15(B) and (C).

(2) Selection of Ampacity. If more than one ampacity is determined from the tables or calculations, the lowest value must be used. There is one exception for adjacent portions of a circuit under certain circumstances.

(B) Tables. Ampacities are noted in the Allowable Ampacities Tables 310.16 through 310.19 and 310.20 through 310.23 for conductors rated 0 to 2000 volts. They can be modified by paragraphs (1) through (6).

(1) General. For explanation of Type Letters, and for recognized size of conductors for the various conductor insulations, see Section 310.13. For installation requirements, see Sections 310.1 through 310.10, and the various articles of this Code. For flexible cords, see Tables 400.4, 400.5(A), and 400.5(B).

(2) Adjustment Factors

(a) More than Three Current-Carrying Conductors in a Raceway or Cable. Where the number of current-carrying conductors in a raceway or cable exceeds three, or where multiconductor cables or single conductors are bundled or stacked longer than 600 mm (24 in.) without proper spacing and not in a raceway, the ampacities shall be reduced as shown in the following table Table 310.15(B)(2)(a). The exceptions to this are listed on the bottom of the table.

TABLE 310.15(B)(2)(a) Adjustment Factors for More than Three Current-Carrying Conductors in a Raceway or Cable

Number of Current-Carrying Conductors	*Percent of Values in Tables 310.16 through 310.19 as Adjusted for Ambient Temperature if Necessary*
4 through 6	80
7 through 9	70
10 through 20	50
21 through 30	45
31 through 40	40
41 and above	35

(FPN): See Appendix B, Table B.310.11 for adjustment factors for more than three current-carrying conductors in a raceway or cable with load diversity.

Exception No. 1. When conductors of different systems, as provided in Section 300.3, are installed in a common raceway or cable the derating factors shown in Table 310.15(B)(2)(a) shall apply to the number of power and lighting (Articles 210, 215, 220, and 230) conductors only.

Exception No. 2. For conductors installed in cable trays, the provisions of Section 392.11 shall apply.

Exception No. 3. Derating factors shall not apply to conductors in nipples having a length not exceeding 610 mm (24 in.).

Exception No. 4. Derating factors shall not apply to underground conductors entering or leaving an outdoor trench if those conductors have physical protection in the form of rigid metal conduit, intermediate metal conduit or rigid nonmetallic conduit having a length not exceeding 3 mm (10 ft.) and the number of conductors does not exceed 4.

Exception No. 5. Adjustment factors shall not apply to Type AC cable or to Type MC cable without an overall outer jacket under the following conditions:

(a) Each cable has not more than three current carrying conductors.

(b) The conductors are 12 AWG copper.

(c) Not more than 20 current carrying conductors are bundled, stacked, or supported on "bridle rings."

A 60 percent adjustment factor shall be applied where the current carrying conductors in these cables that are stacked or bundled longer than 600 mm (24 in.) without maintaining spacing exceeds 20.

(b) More than One Conduit, Tube, or Raceway. Spacing between conduits, tubing, or raceways shall be maintained.

(3) Bare Conductors or Covered Conductors

Where bare or covered conductors are used with insulated conductors, their allowable ampacities shall be limited to that permitted for the adjacent insulated conductors.

(4) Neutral Conductor

(a) A neutral conductor that carries only the unbalanced current from other conductors of the same circuit need not be counted when applying the provisions of Section 310.15(B)(2)(a).

(b) In a three-wire circuit consisting of two-phase wires and the neutral of a four-wire, three-phase wye-connected system, a common conductor carries approximately the same current as the line to neutral load currents of the other conductors and shall be counted when applying the provisions of Section 310.15(B)(2)(a).

(c) On a four-wire, three-phase wye circuit where the major portion of the load consists of nonlinear loads there are harmonic curents present in the neutral conductor and the neutral shall be considered to be a current-carrying conductor.

(5) Grounding or Bonding Conductor

A grounding or bonding conductor shall not be counted when applying the provisions of Section 310.15(B)(2)(a).

(6) 120/240 Volts, Three-Wire, Single-Phase Dwelling Services and Feeders

For dwelling units, conductors, as listed in Table 310.15(B)(6), shall be permitted to be utilized as 120/240-V, three-wire, single-phase service-entrance conductors, service lateral conductors, and feeder conductors that serve as the main power feeder to a dwelling unit and are installed in raceway or cable with or without an equipment grounding conductor. For applications of this section, the main power feeder shall be the feeder(s) between the main disconnect and the lighting and appliance branch circuit panelboard(s). The feeder conductors to a dwelling unit shall not be required to be larger than its service-entrance conductors. The grounded conductor shall be permitted to be smaller than the ungrounded conductors, provided that the requirements of Sections 215.2, 220.22, and 230.42 are met.

TABLE 310.15(B)(6) Conductor Types and Sizes for 120/240-Volt, Three-Wire, Single-Phase Dwelling Services and Feeders

Conductor (AWG or kcmil)		
Copper	*Aluminum or Copper-Clad Aluminum*	*Service or Feeder Ratings (Amperes)*
4	2	100
3	1	110
2	1/0	125
1	2/0	150
1/0	3/0	175
2/0	4/0	200
3/0	250	225
4/0	300	250
250	350	300
350	500	350
400	600	400

TABLE 310.16 Allowable Ampacities of Insulated Conductors, Rated 0 through 2000 Volts, 60° through 90°C (140° through 194°F) Not More than Three Current Conductors in Raceway, Cable, or Earth (Directly Buried)

Based on Ambient Temperature of 30°C (86°F)

Size	*Temperature Rating of Conductor, See Table 310.13*						*Size*
	60°C (140°F)	*75°C (167°F)*	*90°C (194°F)*	*60°C (140°F)*	*75°C (167°F)*	*90 °C (194°F)*	
AWG or kcmil	*Types TW, UF*	*Types RHW, THHW, THW, THWN, XHHW, USE, ZW*	*Types TBS, SA, SIS, FEP, FEPB, MI, RHH, RHW-2, THHN, THHW, THW-2, THWN-2 USE-2, XHH, XHHW, XHHW-2, ZW-2*	*Types TW, UF*	*Types RHW, THHW, THW, THWN, XHHW, USE*	*Types TBS, SA, SIS, THHN, THHW, THW-2, THWN-2, RHH, RHW-2, USE-2, XHH, XHHW, XHHW-2, ZW-2*	*AWG or kcmil*
	Copper			*Aluminum or Copper-Clad Aluminum*			
18	....	....	14	....	....	....	....
16	....	....	18	....	....	....	....
14*	20	20	25	....	....	....	....
12*	25	25	30	20	20	25	12*
10*	30	35	40	25	30	35	10*
8	40	50	55	30	40	45	8

6	55	65	75	40	50	60	6
4	70	85	95	55	65	75	4
3	85	100	110	65	75	85	3
2	95	115	130	75	90	100	2
1	110	130	150	85	100	115	1
1/0	125	150	170	100	120	135	1/0
2/0	145	175	195	115	135	150	2/0
3/0	165	200	225	130	155	175	3/0
4/0	195	230	260	150	180	205	4/0
250	215	255	290	170	205	230	250
300	240	285	320	190	230	255	300
350	260	310	350	210	250	280	350
400	280	335	380	225	270	305	400
500	320	380	430	260	310	350	500
600	355	420	475	285	340	385	600
700	385	460	520	310	375	420	700
750	400	475	535	320	385	435	750
800	410	490	555	330	395	450	800
900	435	520	585	355	425	480	900

TABLE 310.16 (continued)

Size	*Temperature Rating of Conductor, See Table 310.13*						Size
	60°C (140°F)	*75°C (167°F)*	*90°C (194°F)*	*60°C (140°F)*	*75°C (167°F)*	*90 °C (194°F)*	
AWG or kcmil	*Types TW, UF*	*Types RHW, THHW, THW, THWN, XHHW, USE, ZW*	*Types TBS, SA, SIS, FEP, FEPB, MI, RHH, RHW-2, THHN, THHW, THW-2, THWN-2 USE-2, XHH, XHHW, XHHW-2, ZW-2*	*Types TW, UF*	*Types RHW, THHW, THW, THWN, XHHW, USE*	*Types TBS, SA, SIS, THHN, THHW, THW-2, THWN-2, RHH, RHW-2, USE-2, XHH, XHHW, XHHW-2, ZW-2*	*AWG or kcmil*
	Copper			*Aluminum or Copper-Clad Aluminum*			
1000	455	545	615	375	445	500	1000
1250	495	590	665	405	485	545	1250
1500	520	625	705	435	520	585	1500
1750	545	650	735	455	545	615	1750
2000	560	665	750	470	560	630	2000

Correction Factors

Ambient Temp. °C	*For ambient temperatures other than 30°C (86°F), multiply the allowable ampacities shown above by the appropriate factor shown below.*						*Ambient Temp. °F*
21-25	1.08	1.05	1.04	1.08	1.05	1.04	70-77
26-30	1.00	1.00	1.00	1.00	1.00	1.00	78-86
31-35	0.91	0.94	0.96	0.91	0.94	0.96	87-95
36-40	0.82	0.88	0.91	0.82	0.88	0.91	96-104
41-45	0.71	0.82	0.87	0.71	0.82	0.87	105-113
46-50	0.58	0.75	0.82	0.58	0.75	0.82	114-122
51-55	0.41	0.67	0.76	0.41	0.67	0.76	123-131
56-60	....	0.58	0.71	....	0.58	0.71	132-140
61-70	....	0.33	0.58	....	0.33	0.58	141-158
71-80	....	....	0.41	....	....	0.41	159-176

*See Section 240.4(D).

TABLE 310.17 Allowable Ampacities of Single-Insulated Conductors Rated 0 through 2000 Volts, in Free Air

Based on Ambient Temperature of 30°C (86°F)

Size	*Temperature Rating of Conductor, See Table 310.13*						*Size*
	60°C (140°F)	*75°C (167°F)*	*90°C (194°F)*	*60°C (140°F)*	*75°C (167°F)*	*90 °C (194°F)*	
AWG or kcmil	*Types TW, UF*	*Types RHW, THHW, THW, THWN, XHHW, ZW*	*Types TBS, SA, SIS, FEP, FEPB, MI, RHH, RHW-2, THHN, THHW, THW-2, THWN-2 USE-2, XHH, XHHW, XHHW-2, ZW-2*	*Types TW, UF*	*Types RHW, THHW, THW, THWN, XHHW*	*Types TBS, SA, SIS, THHN, THHW, THW-2, THWN-2, RHH, RHW-2, USE-2, XHH, XHHW, XHHW-2, ZW-2*	*AWG or kcmil*
	Copper			*Aluminum or Copper-Clad Aluminum*			
18	....	....	18	....	....	....	....
16	....	....	24	....	....	....	....
14*	25	30	35	....	....	....	....
12*	30	35	40	25	30	35	12*
10*	40	50	55	35	40	40	10*
8	60	70	80	45	55	60	8

6	80	95	105	60	75	80	6
4	105	125	140	80	100	110	4
3	120	145	165	95	115	130	3
2	140	170	190	110	135	150	2
1	165	195	220	130	155	175	1
1/0	195	230	260	150	180	205	1/0
2/0	225	265	300	175	210	235	2/0
3/0	260	310	350	200	240	275	3/0
4/0	300	360	405	235	280	315	4/0
250	340	405	455	265	315	355	250
300	375	445	505	290	350	395	300
350	420	505	570	330	395	445	350
400	455	545	615	355	425	480	400
500	515	620	700	405	485	545	500
600	575	690	780	455	540	615	600
700	630	755	855	500	595	675	700
750	655	785	885	515	620	700	750
800	680	815	920	535	645	725	800
900	730	870	985	580	700	785	900

TABLE 310.17 (continued)

Size	*Temperature Rating of Conductor, See Table 310.13*						*Size*
	60°C (140°F)	*75°C (167°F)*	*90°C (194°F)*	*60°C (140°F)*	*75°C (167°F)*	*90 °C (194°F)*	
AWG or kcmil	*Types TW, UF*	*Types RHW, THHW, THW, THWN, XHHW, ZW*	*Types TBS, SA, SIS, FEP, FEPB, MI, RHH, RHW-2, THHN, THHW, THW-2, THWN-2 USE-2, XHH, XHHW, XHHW-2, ZW-2*	*Types TW, UF*	*Types RHW, THHW, THW, THWN, XHHW*	*Types TBS, SA, SIS, THHN, THHW, THW-2, THWN-2, RHH, RHW-2, USE-2, XHH, XHHW, XHHW-2, ZW-2*	*AWG or kcmil*
	Copper			*Aluminum or Copper-Clad Aluminum*			
1000	780	935	1055	625	750	845	1000
1250	890	1065	1200	710	855	960	1250
1500	980	1175	1325	795	950	1075	1500
1750	1070	1280	1445	875	1050	1185	1750
2000	1155	1385	1560	960	1150	1335	2000

Correction Factors

Ambient Temp. °C	*For ambient temperatures other than 30°C (86°F), multiply the allowable ampacities shown above by the appropriate factor shown below.*						*Ambient Temp. °F*
21-25	1.08	1.05	1.04	1.08	1.05	1.04	70-77
26-30	1.00	1.00	1.00	1.00	1.00	1.00	79-86
31-35	0.91	0.94	0.96	0.91	0.94	0.96	87-95
36-40	0.82	0.88	0.91	0.82	0.88	0.91	96-104
41-45	0.71	0.82	0.87	0.71	0.82	0.87	105-113
46-50	0.58	0.75	0.82	0.58	0.75	0.82	114-122
51-55	0.41	0.67	0.76	0.41	0.67	0.76	123-131
56-60	....	0.58	0.71	....	0.58	0.71	132-140
61-70	....	0.33	0.58	....	0.33	0.58	141-158
71-80	....	....	0.41	....	....	0.41	159-176

*See Section 240.4(D).

TABLE 310.18 Allowable Ampacities of Three Single-Insulated Conductors Rated 0 through 2000 Volts, 150° through 250°C (302° through 482°F). Not More than Three Current-Carrying Conductors in Raceway or Cable

Based on Ambient Air Temperature of 40°C (104°F)

Size	*Temperature Rating of Conductor, See Table 310.13*				*Size*
	150°C (302°F)	*200°C (392°F)*	*250°C (482°F)*	*150°C (302°F)*	
AWG or kcmil	*Type Z*	*Types FEP, FEPB, PFA*	*Types PFAH, TFE*	*Type Z*	*AWG or kcmil*
	Copper		*Nickel or Nickel-Coated Copper*	*Aluminum or Copper-Clad Aluminum*	
14	34	36	39	....	14
12	43	45	54	30	12
10	55	60	73	44	10
8	76	83	93	57	8

6	96	110	117	75	6
4	120	125	148	94	4
3	143	152	166	109	3
2	160	171	191	124	2
1	186	197	215	145	1
1/0	215	229	244	169	1/0
2/0	251	260	273	198	2/0
3/0	288	297	308	227	3/0
4/0	332	346	361	260	4/0

TABLE 310.18 (continued)

Correction Factors					
Ambient Temp. (°C)	*For ambient temperatures other than 40°C (104°F), multiply the allowable ampacities shown above by the appropriate factor shown below*				*Ambient Temp. (°F)*
41-50	0.95	0.97	0.98	0.95	105-122
51-60	0.90	0.94	0.95	0.90	123-140
61-70	0.85	0.90	0.93	0.85	141-158
71-80	0.80	0.87	0.90	0.80	159-176
81-90	0.74	0.83	0.87	0.74	177-194
91-100	0.67	0.79	0.85	0.67	195-212
101-120	0.52	0.71	0.79	0.52	213-248
121-140	0.30	0.61	0.72	0.30	249-284
141-160	....	0.50	0.65	....	285-320
161-180	....	0.35	0.58	....	321-356
181-200	....	....	0.49	....	357-392
201-225	....	....	0.35	....	393-437

TABLE 310.19 Allowable Ampacities of Single-Insulated Conductors Rated 0 through 2000 Volts, 150° through 250°C (302° through 482°F), in Free Air

Based on Ambient Air Temperature of 40°C (104°F)

Size	*Temperature Rating of Conductor, See Table 310.13*				*Size*
	150°C (302°F)	*200°C (392°F)*	*250°C (482°F)*	*150°C (302°F)*	
AWG or kcmil	*Type Z*	*Types FEP, FEPB, PFA*	*Types PFAH, TFE*	*Type Z*	*AWG or kcmil*
	Copper		*Nickel or Nickel-Coated Copper*	*Aluminum or Copper-Clad Aluminum*	
14	46	54	59	....	14
12	60	68	78	47	12
10	80	90	107	63	10
8	106	124	142	83	8

TABLE 310.19 (continued)

Size	*Temperature Rating of Conductor, See Table 310.13*				*Size*
	150°C (302°F)	*200°C (392°F)*	*250°C (482°F)*	*150°C (302°F)*	
AWG or kcmil	*Type Z*	*Types FEP, FEPB, PFA*	*Types PFAH, TFE*	*Type Z*	*AWG or kcmil*
	Copper		*Nickel or Nickel-Coated Copper*	*Aluminum or Copper-Clad Aluminum*	
6	155	165	205	112	6
4	190	220	278	148	4
3	214	252	327	170	3
2	255	293	381	198	2
1	293	344	440	228	1
1/0	339	399	532	263	1/0
2/0	390	467	591	305	2/0
3/0	451	546	708	351	3/0
4/0	529	629	830	411	4/0

Correction Factors

Ambient Temp. (°C)	*For ambient temperatures other than 40°C (104°F), multiply the allowable ampacities shown above by the appropriate factor shown below*				*Ambient Temp. (°F)*
41-50	0.95	0.97	0.98	0.95	105-122
51-60	0.90	0.94	0.95	0.90	123-140
61-70	0.85	0.90	0.93	0.85	141-158
71-80	0.80	0.87	0.90	0.80	159-176
81-90	0.74	0.83	0.87	0.74	177-194
91-100	0.67	0.79	0.85	0.67	195-212
101-120	0.52	0.71	0.79	0.52	213-248
121-140	0.30	0.61	0.72	0.30	249-284
141-160	....	0.50	0.65	....	285-320
161-180	....	0.35	0.58	....	321-356
181-200	....	....	0.49	....	357-392
201-225	....	....	0.35	....	393-437

TABLE 310.20 Ampacities of Not More Than Three Single-Insulated Conductors, Rated 0 through 2000 Volts, Supported on a Messenger, Based on Ambient Air Temperature of 40°C (104°F)

Size	*Temperature Rating of Conductor (See Table 310.13)*				*Size*
	75°C (167°F)	*90°C (194°F)*	*75°C (167°F)*	*90°C (194°F)*	
AWG or kcmil	*TYPES RHW, THHW, THW, THWN, XHHW, ZW*	*TYPES MI, THHN, THHW, THW-2, THWN-2, RHH, RWH-2, USE-2, XHHW, XHHW-2, ZW-2*	*TYPES RHW, THW, THWN, THHW, XHHW*	*TYPES THHN, THHW, RHH, XHHW, RHW-2, XHHW-2, THW-2, THWN-2 USE-2, ZW-2*	*AWG or kcmil*
	Copper		*Aluminum or Copper-Clad Aluminum*		
8	57	66	44	51	8
6	76	89	59	69	6
4	101	117	78	91	4
3	118	138	92	107	3
2	135	158	106	123	2
1	158	185	123	144	1

1/0	183	214	143	167	1/0
2/0	212	247	165	193	2/0
3/0	245	287	192	224	3/0
4/0	287	335	224	262	4/0
250	320	374	251	292	250
300	359	419	282	328	300
350	397	464	312	364	350
400	430	503	339	395	400
500	496	580	392	458	500
600	553	647	440	514	600
700	610	714	488	570	700
750	638	747	512	598	750
800	660	773	532	622	800
900	704	826	572	669	900
1000	748	879	612	716	1000

CORRECTION FACTORS

Ambient Temp. °C	*For ambient temperatures other than 40° (104°F), multiply the ampacities shown above by the appropriate factor shown below.*				*Ambient Temp. °F*
21–25	1.20	1.14	1.20	1.14	70–77
26–30	1.13	1.10	1.13	1.10	79–86
31–35	1.07	1.05	1.07	1.05	88–95
36–40	1.00	1.00	1.00	1.00	97–104
41–45	0.93	0.95	0.93	0.95	106–113
46–50	0.85	0.89	0.85	0.89	115–122
51–55	0.76	0.84	0.76	0.84	124–131
56–60	0.65	0.77	0.65	0.77	133–140
61–70	0.38	0.63	0.38	0.63	142–158
71–80	—	0.45	—	0.45	160–176

TABLE 310.21 Ampacities of Bare or Covered Conductors in Free Air, Based on 40°C (104°F) Ambient, 80°C (176°F) Total Conductor Temperature, 610 mm/sec (2 ft/sec) Wind Velocity

Copper Conductors				*AAC Aluminum Conductors*			
Bare		*Covered*		*Bare*		*Covered*	
AWG or kcmil	*Amperes*	*AWG or kcmil*	*Amperes*	*AWG or kcmil*	*Amperes*	*AWG or kcmil*	*Amperes*
8	98	8	103	8	76	8	80
6	124	6	130	6	96	6	101
4	155	4	163	4	121	4	127
2	209	2	219	2	163	2	171
1/0	282	1/0	297	1/0	220	1/0	231
2/0	329	2/0	344	2/0	255	2/0	268
3/0	382	3/0	401	3/0	297	3/0	312
4/0	444	4/0	466	4/0	346	4/0	364
250	494	250	519	266.8	403	266.8	423
300	556	300	584	336.4	468	336.4	492
500	773	500	812	397.5	522	397.5	548
750	1000	750	1050	477.0	588	477.0	617
1000	1193	1000	1253	556.5	650	556.5	682
—	—	—	—	636.0	709	636.0	744
—	—	—	—	795.0	819	795.0	860
—	—	—	—	954.0	920	—	—
—	—	—	—	1033.5	968	1033.5	1017
—	—	—	—	1272	1103	1272	1201
—	—	—	—	1590	1267	1590	1381
—	—	—	—	2000	1454	2000	1527

(C) Engineering Supervision.

With engineering supervision the following formula can be used to calculate ampacities:

$$I = \sqrt{\frac{TC - (TA + DELTA\ TD)}{RDC(1 + YC)RCA}}$$

where

TC = conductor temperature in degrees Celsius (°C)
TA = ambient temperature in degrees Celsius (°C)
DELTA TD = dielectric loss temperature rise
RDC = dc resistance of conductor at temperature TC
YC = component ac resistance resulting from skin effect and proximity effect
RCA = effective thermal resistance between conductor and surrounding ambient

FPN: See Appendix B for examples of formula application.

310.60 Conductors Rated 2001 to 35,000 Volts.

(A) Definitions.

(B) Ampacities of Conductors Rated 2001 to 35,000 Volts.

(C) Tables.

ARTICLE 312 CABINETS, CUTOUT BOXES, AND METAL SOCKET ENCLOSURES

312.1 Scope

I. Installation

312.2 Damp, Wet, or Hazardous (Classified) Locations

When installed in wet locations, they must be weatherproof. In damp or wet locations, surface boxes must prevent moisture from entering and have at least 6 mm (¼ in.) of air space between it and the wall or support. Nonmetallic enclosures can be installed on concrete, tile, or masonry without an airspace. In hazardous locations conformance must be made with the requirements of Articles 500 through 517.

312.3 Position in Wall

When installed in noncombustible material, they may be set back 6 mm (¼ in.) When installed in combustible material, they must be flush with the surface.

312.5 Cabinets, Cutout Boxes, and Meter Socket Enclosures

The opening must be closed. When metal cabinets, boxes, or enclosures are used with open wiring or knob-and-tube work, the conductors must come in through an insulated bushing. In dry areas, they may enter through flexible tubing. Cables must be secured to the box. There is an exception.

312.6 Deflection of Conductors

Refer to Tables 312.6(A) through (C) in the *NEC*® for minimum width of gutters and bending spaces.

312.7 Space in Enclosures

312.8 Enclosures for Switches or Overcurrent Devices

312.9 Side or Back Wiring Spaces or Gutters

II. Construction Specifications

312.10 Material

If metal, it must have corrosion protection. Sheet steel cabinets must be at least 1.35 mm (0.053 in.) thick uncoated. Nonmetallic cabinets must be approved by the authorities.

312.11 Spacing

ARTICLE 314
OUTLET, DEVICE, PULL AND JUNCTION BOXES; CONDUIT BODIES; FITTINGS; AND MANHOLES

I. Scope and General

314.1 Scope

Refer to this section in the *NEC*® for inclusions and exclusions.

314.2 Round Boxes

These are not permitted if locknuts or bushings are required on the side of the box.

314.3 Nonmetallic Boxes

These are only to be used with nonmetallic sheathed cable or raceways, flexible cords, open wire on insulator installations, or knob-and-tube concealed wiring. They can be used with metal raceway or metal-jacketed cable only if there is a means for integral bonding or internal bonding. There are two exceptions.

314.4 Metal Boxes

These must be grounded as provided in Article 250.

314.5 Short Radius Conduit Bodies

Conduit bodies, like capped elbows, which enclose 6 AWG or smaller conductors and are used only to allow the installation of the raceway and its conductors can not have splices, taps, or devices and they must be large enough for free space for the conductors.

II. Installation

314.15 Damp, Wet Locations or Hazardous (Classified) Locations

Moisture must be prevented from getting into the box. They must be listed for use in wet locations. When installed in a hazardous location, conform to Articles 500 through 517.

314.16 Number of Conductors in Outlet, Device and Junction Boxes, and Conduit Bodies

The reader is referred to this section for complete information. These items must be large enough to allow free space for all the enclosed conductors. The volume of the box as calculated in this section cannot be smaller than the fill calculation. This does not apply to terminal housings supplied with the motors (refer to Section 430.12). For enclosed conductors 4 AWG or larger, the provisions of Section 314.28 must be complied with.

(A) Box Volume Calculations. The volume is the total of the assembled sections and if used, spaces from plaster rings, domed covers, extension rings, and the like that are made from boxes listed in Table 314.16(A) or have the volume marked.

(1) Standard Boxes. Given in Table 314.16(A).
(2) Other Boxes. If the box is 100 cubic inches or less [except those in Table 314.16(A)] and nonmetallic boxes must have the capacity marked by the manufacturer. If the box is listed in Table 314.16(A) and has a larger capacity, the manufacturer can mark it.

(b) Box Fill Calculations.

(1) Conductor Fill. Every conductor that originates outside the box and terminates or is spliced in the box and each conductor passing through the box is counted once. The fill is computed using Table 314.16(B). If a conductor has no part leaving the box, it is not counted. There is an exception for equipment grounding conductors.
(2) Clamp Fill. If an internal clamp is used there, a single volume allowance in accordance with Table 314.16(B) is made based on the largest conductor in the box.
(3) Support Fittings Fill. If a stud or hickey is there, a single volume allowance in accordance with Table 314.16(B) is made based on the largest conductor in the box.
(4) Device or Equipment Fill. For a strap or yoke containing a device or equipment, a double volume allowance is made in accordance with Table 314.16(B) based on the largest conductor connected to the device or equipment.
(5) Equipment Grounding Fill. If equipment grounding conductors come into the box, a single volume allowance is made in accordance with Table 314.16(B) and is based on the largest equipment grounding conductor present. If additional equipment grounding conductors as noted in 250.146(D), come into the box an additional volume allowance is made based on the largest equipment grounding conductor in this set.

(c) Conduit Bodies. Generally, they must have cross-sectional area not less than twice the area of the largest conduit or tubing attached. The maximum number of conductors is shown in Table 1, Chapter 9. There are additional requirements noted in the Code.

314.17 Conductors Entering Boxes, Conduit Bodies, or Fittings

All openings must be closed. All conductors coming into a box, conduit body, or fitting must be protected against abrasion. Special requirements are indicated with open wiring or concealed knob-and-tube wiring. If cable or raceway is used, the cable or raceway must be fastened to the box, conduit body, or fitting. Refer to Section 300.15(C) of the *NEC*® for all rules and to Section 300.4(F) for wire 4 AWG or larger. The reader is referred to this section for complete details.

314.19 Boxes Enclosing Flush Devices

They must be closed on the back and sides and a support for the device (usually a stud or screw hole) must be in the box. The screw that supports the box cannot be used to attach the device.

314.20 In Wall or Ceiling

When the box is set in concrete, tile, or other noncombustible material, it can be set back as much as 6 mm (¼ in.) When it is set in combustible material, it has to be flush with the finished surface.

314.21 Repairing Plaster and Drywall or Plasterboard

314.22 Exposed Surface Extensions

314.23 Supports

This section details methods of support. It indicates where they are needed and what can be used.

314.24 Depth of Outlet Boxes

314.25 Covers and Canopies

314.27 Outlet Boxes

They must be designed for the purpose to be used (i.e., floor or lighting fixture). A fixture support must be in a lighting fixture box. Cannot be used as sole support for ceiling paddle fans unless listed for that purpose. There are special cases. The reader is referred to this section in the *NEC*®.

314.28 Pull and Junction Boxes and Conduit Bodies

(A) Minimum Size. The following rules apply only when the wire is 4 AWG or larger. In a straight pull, the box must be at

TABLE 314.16(A) Metal Boxes

Box Trade Size			*Minimum Volume*		*Maximum Number of Conductors**						
mm	*in.*		*cm³*	*in.³*	*18*	*16*	*14*	*12*	*10*	*8*	*6*
100 × 32	(4 × 1¼)	round/octagonal	205	12.5	8	7	6	5	5	5	2
100 × 38	(4 × 1½)	round/octagonal	254	15.5	10	8	7	6	6	5	3
100 × 54	(4 × 2⅛)	round/octagonal	353	21.5	14	12	10	9	8	7	4
100 × 32	(4 × 1¼)	square	395	18.0	12	10	9	8	7	6	3
100 × 38	(4 × 1½)	square	344	21.0	14	12	10	9	8	7	4
100 × 54	(4 × 2⅛)	square	497	30.3	20	17	15	13	12	10	6
120 × 32	(4¹¹⁄₁₆ × 1¼)	square	418	25.5	17	14	12	11	10	8	5
120 × 38	(4¹¹⁄₁₆ × 1½)	square	484	29.5	19	16	14	13	11	9	5
120 × 54	(4¹¹⁄₁₆ × 2⅛)	square	689	42.0	28	24	21	18	16	14	8
75 × 50 × 38	(3 × 2 × 1½)	device	123	7.5	5	4	3	3	3	2	1
75 × 50 × 50	(3 × 2 × 2)	device	164	10.0	6	5	5	4	4	3	2
75 × 50 × 57	(3 × 2 × 2¼)	device	172	10.5	7	6	5	4	4	3	2

75 × 50 × 65	(3 × 2 × 2½)	device	205	12.5	8	7	6	5	5	4	2
75 × 50 × 70	(3 × 2 × 2¾)	device	230	14.0	9	8	7	6	5	4	2
75 × 50 × 90	(3 × 2 × 3½)	device	295	18.0	12	10	9	8	7	6	3
100 × 54 × 38	(4 × 2⅛ × 1½)	device	169	10.3	6	5	5	4	4	3	2
100 × 54 × 48	(4 × 2⅛ × 1⅞)	device	213	13.0	8	7	6	5	5	4	2
100 × 54 × 54	(4 × 2⅛ × 2⅛)	device	238	14.5	9	8	7	6	5	4	2
95 × 50 × 65	(3¾ × 2 × 2½)	masonry box/gang	230	14.0	9	8	7	6	5	4	2
95 × 50 × 90	(3¾ × 2 × 3½)	masonry box/gang	344	21.0	14	12	10	9	8	7	2
min. 44.5 depth	FS—single cover/gang (1¾)		221	13.5	9	7	6	6	5	4	2
min. 60.3 depth	FD—single cover/gang (2⅜)		295	18.0	12	10	9	8	7	6	3
min. 44.5 depth	FS—multiple cover/gang (1¾)		295	18.0	12	10	9	8	7	6	3
min. 60.3 depth	FD—multiple cover/gang (2⅜)		395	24.0	16	13	12	10	9	8	4

*Where no volume allowances are required by 370.16(B)(2) through 370.16(B)(5).

TABLE 314.16(B) Volume Allowance Required per Conductor

Size of Conductor (AWG)	*Free Space within Box for Each Conductor*	
	cm^3	*m^3*
18	24.6	1.50
16	28.7	1.75
14	32.8	2.00
12	36.9	2.25
10	41.0	2.50
8	49.2	3.00
6	81.9	5.00

least 8 times the metric/designator (trade size) of the largest raceway in length. When angle pulls or U-pulls are used, the distance between the raceway entrances and the opposite wall must be at least 6 times the metric designator (trade size) of the largest raceway in a row. In addition, that distance must be increased by the diameter of each additional raceway in any one row on the same wall of the box. Since each row is calculated individually, the largest distance for a single row is used. Two raceways that enclose the same conductor cannot be closer than 6 times the metric designator (trade size) of the larger raceway. There is an exception where the entry is opposite a removable cover. There is also a provision for boxes or conduit bodies whose dimensions are less than calculated where combination of conductors are used.

(B) Conductors in Pull or Junction Boxes

(C) Covers

(D) Permanent Barriers. If a permanent barrier is in a box, each section is considered a box.

314.29 Boxes and Conduit Bodies to Be Accessible

III. Construction Specifications

314.40 Metal Boxes, Conduit Bodies, and Fittings

314.41 Covers

314.42 Bushings

314.43 Nonmetallic Boxes

314.44 Marking

IV. Manholes and Other Electric Enclosures Intended for Personnel Entry

314.50 General

314.51 Strength

314.52 Cabling Work Space

314.53 Equipment Work Space

314.54 Bending Space for Conductors

314.55 Accesss to Manholes

314.56 Access to Vaults and Tunnels

314.57 Ventilation

314.58 Guarding

314.59 Fixed Ladders

V. Pull and Junction Boxes for Use on Systems Over 600 Volts, Nominal

314.70 General

314.71 Size of Pull and Junction Boxes

314.72 Construction and Installation Requirements

ARTICLE 320
ARMORED CABLE
Type AC

I. General

320.1 Scope

320.2 Definition

This type of cable is composed of insulated conductors in a flexible metal enclosure. Refer to 320.100.

II. Installation

320.10 Uses Permitted

It can be used exposed or concealed and in cable trays where so identified. It can not be subject to physical damage. It is permitted in dry locations and in plaster finish on brick or other masonry except in damp or wet locations. It can also be used in the voids of masonry block or tile walls when the walls are not exposed, below grade or subject to excessive moisture or dampness.

320.12 Uses Not Permitted

It cannot be used where prohibited in the Code. This includes in theaters or similar locations except as noted in Article 518.4; motion picture studios; in hazardous locations except where noted in Sections 501.4(B) Exception, 502.4(B) Exception No. 1, and 504.20; where exposed to corrosive fumes; in storage battery rooms; in hoistways or on elevators except as noted in Section 620.21; or in commercial garages where noted in Article 511.4 and 511.7.

320.15 Exposed Work

320.17 Through or Parallel to Framing Members

320.23 In Accessible Attics

320.24 Bending Radius

They must be made so that the cable will not be damaged and the radius of the curve of the inner edge cannot be less than five times the diameter of the cable.

320.30 Securing and Supporting

320.40 Boxes and Fittings

320.80 Ampacity

III. Construction Specifications

320.150 Construction

The cable consists of insulated conductors enclosed in a flexible metal tape armor with a copper or aluminum bonding strip inside and in contact with the armor.

320.104 Conductors

320.108 Grounding

320.120 Marking

ARTICLE 322
FLAT CABLE ASSEMBLIES
Type FC

I. General

322.1 Scope

322.2 Definition

II. Installation

322.10 Uses Permitted

322.12 Uses not Permitted

322.30 Securing and Supporting

322.40 Boxes and Fittings

322.56 Splices and Taps

III. Construction

322.100 Construction

322.104 Conductors

322.112 Insulation

322.120 Marking

ARTICLE 324
FLAT CONDUCTOR CABLE
Type FCC

I. General

324.1 Scope

The wiring covered in this article is a field-installed system to be used for branch circuits under carpet squares.

324.2 Definitions

The reader is referred to the Code for these definitions.

II. Installation

324.10 Uses Permitted

It can be used for general-purpose, appliance, and individual branch circuits; on floors that are hard, sound, or smooth made of materials like or similar to concrete, ceramic, composition, and wood; on walls in surface metal raceways; in damp locations; and on heated floors where the temperature is higher than 86°F if approved for those temperatures.

The voltage ratings are up to 150 V to ground and up to 300 V between ungrounded conductors. The current ratings are up to 20 A for general-purpose and appliance branch circuits and up to 30 A for individual branch circuits.

When on a floor, it must be covered with carpet squares. Release-type adhesive only is permitted. If it exceeds 2.3 mm (0.09 in.) above floor level it must be feathered or tapered at the edges to floor level. There are other requirements.

324.12 Uses Not Permitted

It cannot be used outdoors, in wet locations, in classified hazardous locations, where there are corrosive vapors present, or in residential, school, or hospital buildings.

324.18 Crossings

324.30 Securing and Supporting

324.40 Boxes and Fittings

324.42 Devices

324.56 Splices and Taps

324.60 Grounding

III. Construction

324.100 Construction

324.112 Insulation

324.180 Markings

ARTICLE 326
INTEGRATED GAS SPACER CABLE
Type IGS

I. General

326.1 Scope

326.2 Definition

II. Installation

326.10 Uses Permitted

326.12 Uses Not Permitted

326.24 Bending Radius

326.26 Bends

326.40 Fittings

326.80 Ampacity

III. Construction Specifications

326.104 Conductors

326.112 Insulation

326.116 Conduit

326.120 Marking

ARTICLE 328
MEDIUM-VOLTAGE CABLE
Type MV

I. General

328.1 Scope

328.2 Definition

This cable is rated above 2000 V and is a single or multiconductor solid dielectric insulated cable.

II. Installation

328.10 Uses Permitted

It can be used in dry or wet locations, in raceways, directly buried, and in messenger supported cable in systems up to 35,000 V. Refer to Sections 392.3(B)(1) and 300.50.

328.12 Uses Not Permitted

It cannot be used in cable trays or direct sunlight unless identified for that purpose.

328.80 Ampacity

III. Construction Specifications

328.100 Construction

ARTICLE 330
METAL-CLAD CABLE
Type MC

I. General

330.1 Scope

330.2 Definition

Metal-clad cable is Type MC. It is a cable in which the conductors are enclosed in a corrugated metal sheath or interlocking metal tape, insulated, and the entire cable is put together at the factory.

II. Installation

330.10 Uses Permitted

This cable can be used for almost all types of applications, including wet locations, provided that it will not sustain physical damage and is not prohibited in another part of the *NEC*®. A listing is given in the *NEC*® for both general uses and specific uses. In wet locations one of the following must be adhered to: moisture will not get through the metal cover, a lead sheath that will not let moisture through is put under the metal cover, or the insulated conductors enclosed by the metal cover are listed for a wet location. Refer to this section in the Code.

Type MC cable must be installed in accordance with the requirements of Articles 300, 490, 725, and Section 770.52. When installed in cable tray, direct buried, as service-entrance cable outside buildings or as aerial cable, you must refer to the appropriate articles in the *NEC*®.

330.12 Uses Not Permitted

This cable cannot be used in a location where it is subject to corrosive conditions that are destructive. The *NEC*® gives as examples direct buried cable in earth or concrete, under fills, chlorides, caustic alkalis, and the like.

330.17 Through or Parallel to Framing Members

It must comply with 300.4.

330.23 In Accessible Attics

Must comply with 320.23.

330.24 Bending Radius

The following are the allowable radii of the curve of the inner edge: for shielded conductors, use the greater of either 12 times the diameter of one conductor or 7 times the overall diameter of the cable; for interlocked armor or corrugated sheath, 7 times the diameter of the sheath; for smooth sheath, use the following:

> Cable up to 19 mm (¾ in.) diameter: 10 times diameter
> Cable between 19 mm (¾) and 38 mm (1½ in.) diameter: 12 times diameter
> Cable larger than 38 mm (1½ in.) diameter: 15 times diameter

330.30 Securing and Supporting

There are requirements for supported and unsupported cables.

330.31 Single Conductors

330.40 Boxes and Fittings

330.80 Ampacity

Refer to 310.15 or 310.60 for 14 AWG and larger conductors. Use Table 402.5 for 18-16 AWG conductors. There are additional requirements for cable installed in cable tray and cable grouped together supported by messenger cable.

C. Construction Specifications

330.104 Conductors

Copper, aluminum, or copper-clad aluminum must be used, either stranded or solid. The smallest sizes allowed are 18 AWG for copper and 12 AWG for aluminum or copper-clad aluminum.

330.108 Equipment Grounding

330.112 Insulation

330.116 Sheath

The sheath must be either smooth, welded, and corrugated or interlocking.

ARTICLE 332
MINERAL-INSULATED, METAL-SHEATHED CABLE
Type MI

I. General

332.1 Scope

332.2 Definition

Type MI cable is factory assembled with one or more conductors using a highly compressed refractory mineral insulation. The entire assembly is covered with a gastight and liquidtight continuous copper or alloy steel sheath.

II. Installation

332.10 Uses Permitted

It can be used for services, feeders, and branch circuits; power, control, lighting, and signal circuits; wet, dry, or continuously moist locations; exposed or concealed; indoors and outdoors; embedded in concrete, plaster, fill, and other masonry; hazardous locations; underground when protected; corrosive locations that will not harm the sheath; and where exposed to gasoline and oil.

332.12 Uses Not Permitted

Where exposed to corrosive materials that will harm the sheath unless it is properly protected.

332.17 Through or Parallel to Framing Members

332.24 Bending Radius

332.30 Securing and Supporting

332.31 Single Conductors

332.40 Boxes and Fittings

332.80 Ampacity

III. Construction Specifications

332.104 Conductors

Solid copper or nickel-clad copper.

332.108 Equipment Grouding

332.112 Insulation

Highly compressed refractory mineral.

332.116 Sheath

Continuous construction that ensures mechanical protection and a moisture seal.

ARTICLE 334
NONMETALLIC-SHEATHED CABLE
Types NM, NMC, and NMS

I. General

334.1 Scope

334.2 Definition

Nonmetallic-sheathed cable is a cable in which the insulated conductors are enclosed in a nonmetallic material that is moisture-resistant and flame-retardant. The cable is put together at the factory.

334.6 Listed

II. Installation

334.10 Uses Permitted

These types of cable can be used in one- and two-family dwellings, multifamily dwellings of types III, IV, and V construction except as noted in Section 334.12, and cable trays when they are noted for this use. These cables can also be used in other structures of Types III, IV, and V construction with limitations. Type NM cable can be used both in dry locations exposed and concealed as well as being fished in air voids of block and tile if it is not damp or moist. Type NMC cable can be used exposed or concealed in damp, dry, or corrosive locations, inside and outside of block walls, and in shallow chases protected against nails or screws. Type NMS can be used in the same locations as Type NM. The reader is referred to this section of the *NEC*® for a complete list of locations where these cables can be used.

334.12 Uses Not Permitted

Type NM, NMC, and NMS cannot be used as service-entrance cable, in other than one, two, and multifamily dwellings as open runs in dropped or suspended ceilings, in commercial garages, in motion picture studios, in theaters, and the like unless otherwise provided in the *NEC*®, in hoistways, storage battery rooms, hazardous locations, and in poured concrete cement or aggregate. There are additional limitations for Types NM and NMS cable. The reader is referred to this section of the *NEC*® for a complete list of locations where types NM, NMC, and NMS are not permitted.

334.15 Exposed Work

334.17 Through or Parallel to Framing Members

334.23 In Accessible Attics

334.24 Bending Radius

334.30 Securing and Supporting

The cable must be secured at least every 1.4 m (4½ ft.) and not more than 300 mm (12 in.) from a fitting, box, or cabinet. Two exceptions are noted. Only certain supports are approved.

334.40 Boxes and Fittings

334.80 Ampacity

III. Construction Specifications

334.100 Construction

334.104 Conductors

334.108 Equipment Grounding

334.112 Insulation

334.116 Sheath

ARTICLE 336
POWER AND CONTROL TRAY CABLE
Type TC

I. General

336.1 Scope

336.2 Definition

This cable is a nonmetallic-sheathed cable used in cable trays, raceways, or supported by messenger wire with insulated conductors. The grounding conductor can be either bare or covered.

II. Installation

336.10 Uses Permitted

This cable can be used for power, lighting, control, signal, and communications circuits; in cable trays, raceways, or supported by messenger wire; in hazardous locations when permitted in Articles 318, 501, 502, 504, and 505; under certain conditions where permitted by Article 725 for Class I circuits; for nonpower-limited fire alarm circuits in accordance with Section 760.27; and for industrial establishments with qualified supervision and maintenance. There are additional requirements for support, identification, and equipment grounding.

336.12 Uses Not Permitted

Type TC cable cannot be used where exposed to physical damage; on open brackets except as permitted in 340.10(6); where exposed to direct sunlight unless sunlight resistant; and direct buried unless approved for that purpose.

336.24 Bending Radius

336.80 Ampacity

III. Construction Specifications

336.100 Construction

A metallic sheath cannot be put over or under the nonmetallic sheath. Metallic sheaths can be put over groups of conductors, under the outer jacket or both.

336.104 Conductors

Minimum-size conductors are 18 AWG for copper and 12 AWG for aluminum and copper-clad aluminum. Maximum size is 1000 kcmil. 14 AWG copper and larger and 12 AWG or larger aluminum or copper-clad aluminum are to be of the types shown in Table 310.13 or 310.62.

(A) Fire Alarm Systems. It must be installed as noted in Section 760.27 for fire alarms systems.

(B) Thermocouple Circuits. It must be installed as noted in Article 725. There are additional requirements.

(C) Class 1 Circuit Conductors. When insulated conductors of sizes 18 AWG or 16 AWG copper are used, they shall be as noted in Section 725.27.

336.116 Jacket

The outer jacket must be a flame-retardant, nonmetallic material.

336.120 Marking

ARTICLE 338
SERVICE-ENTRANCE CABLE
Types SE and USE

I. General

338.1 Scope

338.2 Definition

This cable is either Type SE or USE and is used primarily for services. Type SE has a flame-retardant and moisture-resistant cover. Type USE, which is the one used underground, has a moisture-resistant cover.

II. Installation

338.10 Uses Permitted

(A) Service-Entrance Conductors. When used as service-entrance conductors, you must follow Article 230. If protected as noted in Section 300.5(D), Type USE can come above ground outside in meter enclosures or terminations when used as a service lateral.

(B) Branch Circuits or Feeders.

(1) Grounded Conductor Insulated. Type SE can be used if all conductors are rubber-covered or thermoplastic type.

(2) Grounded Conductor Not Insulated. Generally, Type SE cable with a noninsulated grounded conductor cannot be used for branch circuit or feeder wiring in a building. There is an exception.

(3) Temperature Limitations. Do not exceed temperature specified.

(4) Installation Methods for Branch Circuits and Feeders.

Review applicable sections of the *NEC*®.

338.24 Bending Radius

III. Construction

338.100 Construction

338.120 Marking

ARTICLE 340
UNDERGROUND FEEDER AND BRANCH-CIRCUIT CABLE
Type UF

I. General

340.1 Scope

340.2 Definition

A listed factory assembly with an integral covering of nonmetallic material for direct burial in the earth.

II. Installation

340.10 Uses Permitted

In underground, including direct burial, as a feeder or branch circuit as long as the overcurrent requirements of Section 240.3 are adhered to. It can be used for interior wiring in wet, dry, or corrosive areas provided that all other conditions are met. In certain of these cases, as described in Section 690.31, it can be used for solar photovoltaic systems. Refer to Section 300.5 for underground requirements. When single conductor cables are used, all cables of the feeder or branch circuit have to be run in the same trench or wireway. Single conductor cables can be nonheating leads for heating cable. If supported by cable trays, this cable must be multiconductor. If installed in accordance with Article 334 it can be used as nonmetallic sheathed cable.

340.12 Uses Not Permitted

This type of cable cannot be used for service entrances, in commercial garages, in theaters, in storage battery rooms, in motion picture studios, in hazardous locations, in hoistways, in poured cement, where exposed to direct rays of sun unless approved for sunlight installation, where subject to physical damage, or as an overhead cable except as messanger supported wiring.

340.24 Bending Radius

340.80 Ampacity

Defer to 310.15, 60°C (140°F) conductors.

III. Construction Specifications

340.104 Conductors

This cable is Type UF. Sizes allowed are minimum 14 AWG copper or 12 AWG aluminum or copper-clad aluminum and maximum 4/0 AWG.

340.108 Equipment Grounding

The cable can have an insulated or bare conductor only for equipment grounding purposes.

340.112 Insulation

Moisture resistant type found in Table 310.13

340.116 Sheath

Flame retardant; moisture, fugus, and corrosion resistant, suitable for direct burial in earth.

ARTICLE 342
INTERMEDIATE METAL CONDUIT
Type IMC

I. General

342.1 Scope

342.2 Definition

This conduit is a circular metal raceway threadable and designed for the protection and routing of conductors and cable. It can be used as an equipment grounding conductor with its couplings and fittings.

342.6 Listing Requirements

IMC, the factory elbows, couplings, and fittings must be listed.

II. Installation

342.10 Uses Permitted

Intermediate metal conduit can be used in most places and under most conditions. It can be used in the earth or in corrosive locations if it has corrosion protection. It can be used in cinder fill only if 50 mm (2 in.) of noncinder concrete is in direct contact, the conduit is 450 mm (18 in.) or more under the fill, or there is proper corrosion protection. In wet locations all supports, belts, screws, etc. must be protected against corrosion.

342.14 Dissimilar Metals

With the exception of aluminum fittings and enclosures, dissimilar metals in contact must not be used where possible.

342.20 Size

The conduit used cannot be smaller than metric designator 16 (trade size ½) or larger than metric designator 103 (trade size 4). See 300.1(C) for metric designators.

342.22 Number of Conductors

Refer to Chapter 9, Tables 1 and 4.

342.24 Bends—How Made

The conduit must not be injured. Refer to Table 344.24 for the radius of the curve for field bends.

342.26 Bends—Number in One Run

The maximum that is permitted is not more than 360° in bends between pull points.

342.28 Reaming and Threading

All rough edges must be removed when the conduit is cut. When threaded in the field, a tapered cutting die with a taper of 1 in 16 (¾ in. taper per foot) must be used.

342.30 Securing and Supporting

(A) Securely Fastened. It must be securely fastened within 900 mm (3 ft.) of all boxes, cabinets, conduits, or other conduit terminations. When structural members interfere with this, then the distance can be increased to 1.5 m (5 ft.). If approved, it does not have to be fastened within 900 mm (3 ft.) of a service head above a roof.

(B) Supports. It has to be supported at intervals no longer than 3 m (10 ft.) Table 344.30(B)(2) indicates the spacing of supports for straight runs of conduit with the proper couplings and there is a prevention of the transmission of stresses. Exposed vertical risers can be supported at intervals of 6 m

(20 ft.), if the couplings are threaded, firmly supported at top and bottom, and there is no other intermediate support available. Horizontal runs that are supported by framing members not longer than 3 m (10 ft.) apart with supports within 900 mm (3 ft.) of the terminations are permitted.

342.42 Couplings and Connectors

Threadless couplings and connectors can be used. Running threads cannot be used. Use concrete type in concrete and raintight in wet areas.

342.46 Bushings

A bushing must be used at the point where the conduit goes into a box, fitting, or other enclosure. This is to protect the wire. An exception is made if protection is provided with the box or fitting.

342.56 Splices and Taps

They are permitted only if made in accordance with Section 300.15.

342.60 Grounding

III. Construction Specifications

342.120 Marking

The conduit must be marked every 1.5 m (5 ft.) with the letter IMC and as required by 110.24.

342.130 Standard Lengths

This type of conduit has a standard length of 3 m (10 ft.) with one coupling and both ends threaded. Shorter or longer lengths with or without couplings or threads are permitted.

ARTICLE 344
RIGID METAL CONDUIT
Type RMC

I. General

344.1 Scope

344.2 Definition

Rigid metal conduit is a raceway that is listed, has a cross section that is circular, has integral or associated couplings, is used with listed fittings to give electrical continuity, and is approved for the installation of electrical conductors.

344.6 Listing Requirements

This type of conduit, associated fittings, factory elbows and couplings must be listed.

II. Installation

344.10 Uses Permitted

(A) All Atmospheric Conditions and Occupancies. It is permitted in all atmospheric conditions and all occupancies. If the conduit's protection against corrosion is only enamel, it can be used only indoors and in areas that do not have severe corrosive problems. Aluminum enclosures and fittings can be used with steel conduit and vice versa.

(B) Corrosion Environments. If the conduit is protected from corrosion and is suitable, the conduit, elbows, couplings, and fittings can be installed in concrete, in areas where there is a severe corrosive problem, and in direct contact with the earth.

(C) Cinder Fill. It cannot be used in or under cinder fill when there may be permanent moisture unless it is suitable for this type of installation, there are 50 mm (2 in.) or more of noncinder concrete all around it, it is at least 450 mm (18 in.) under the fill, or where it is suitable for the condition and has corrosion protection.

(D) Wet Locations. Everything must be corrosion-resistant (bolts, straps, etc.).

344.14 Dissimilar Metals

Contact between dissimilar metals must be avoided where possible to eliminate galvanic action. Aluminum fittings are permitted with steel and vice versa where not subject to severe corrosion.

344.20 Size

Metric designator 16 (trade size ½) is the minimum size permitted unless used for enclosing motor leads [see Section 430.145(B)]. The maximum size permitted is metric designator 155 (trade size 6). See 300.1(C) for metric designators.

344.22 Number of Conductors in Conduit

Refer to Chapter 9, Table 1, using the conduit dimensions of Table 4, Chapter 9.

344.24 Bends—How Made

The conduit must not be damaged and the internal diameter will not be reduced. Refer to Table 344.24 for the radius of the inner edge for field bends.

344.26 Bends—Number in One Run

The maximum that is permitted is not more than 360° in bends between pull points.

344.28 Reaming and Threading

All rough edges must be removed when the conduit is cut. When threaded in the field, a tapered cutting die 1 in 16 taper (¾ in. taper per foot) must be used.

344.30 Securing and Supporting

The conduit must be supported at intervals no longer than 3 m (10 ft.) and fastened not more than 900 mm (3 ft.) from a box, fitting, etc. It must be installed as a complete system. Fastening can be increased to 1.5 m (5 ft.) where structural members do not per-

TABLE 344.24 Radius of Conduit Bends

Conduit (Size)		*One Shot and Full Shoe Benders*		*Other Bends*	
Metric Designator	*Trade Size*	*mm*	*in.*	*mm*	*in.*
16	½	101.6	4	101.6	4
21	¾	114.3	4½	127	5
27	1	146.05	5¾	152.4	6
35	1¼	184.15	7¼	203.2	8
41	1½	209.55	8¼	254	10
53	2	241.3	9½	304.8	12
63	2½	266.7	10½	381	15
78	3	330.2	13	457.2	18
91	3½	381	15	533.4	21
103	4	406.4	16	609.6	24
129	5	609.6	24	762	30
155	6	762	30	914.4	36

mit fastening within 900 mm (3 ft.). For supports for straight runs with threaded couplings, refer to Table 344.30(B)(2). No stress is to be transmitted to the terminations. Vertical risers from items like machine tools can have the distance between supports at 6 m (20 ft.) if the conduit uses threaded couplings and is properly supported at the top and bottom. Where approved, the conduit does not have to be fastened within 900 mm (3 ft.) of a service head for a termination of a mast above the roof. Horizontal runs that are supported by openings through framing members not less than 3 m (10 ft.) and fastened within 900 mm (3 ft.) of termination points are permitted.

344.42 Couplings and Connectors

Threadless couplings and connectors can be used. Running threads cannot be used. Use concrete type in concrete and raintight in wet areas.

344.46 Bushings

A bushing must be used at the point where the conduit goes into a box, fitting, or other enclosure. This is to protect the wire. An exception is made if protection is provided with the box or fitting.

344.56 Splices and Taps

Refer to Article 300.15.

344.60 Grounding

It is permitted as an equipment grounding conductor.

III. Construction Specifications

344.120 Marking

If the conduit is nonferrous of corrosion-resistant material, it must be marked to indicate it. At every 3 m (10 ft.) the conduit must be marked with identification (refer to Section 110.21).

344.130 Standard Lengths

This type of conduit has a standard in length of 3 m (10 ft.) including one attached coupling. It has to be threaded at each end. Smaller and larger lengths can be used with or without couplings and threads.

ARTICLE 348
FLEXIBLE METAL CONDUIT
Type FMC

I. General

348.1 Scope

348.2 Definition

This is a circular raceway of helically wound, formed, interlocked metal strip.

348.6 Listing Requirements

Must be listed.

II. Installation

348.10 Uses Permitted

It can be used in exposed and concealed locations.

348.12 Uses Not Permitted

It cannot be used in wet locations unless approved for this and the liquid does not get into the conduit, in hoistways unless permitted by Section 620.21(A)(1), in storage battery rooms, in hazardous locations unless allowed by Sections 501.4(B) and 504.20, where materials might deteriorate conductors, underground or in poured concrete or aggregate, and where subject to physical damage.

348.20 Size

Minimum size permitted is metric designator 12 (trade size ⅜). The use of metric designator 12 (⅜) is allowed for leads to motors [Section 430.145(b)], not longer than 1.8 m (6 ft.) as part of a listed assembly or taps to lighting fixtures [Section 410.67(C)] or utilization equipment, for manufactured wiring systems [Section 604.6(A)], in hoistways Section 620.21, and as permitted by Section 410.77(C) as part of a listed assembly to connect wired fixture sections. There are five exceptions. The maximum size permitted is metric designator 103 (trade size 4).

348.22 Number of Conductors

Cannot exceed the percentage fill of Table 1, Chapter 9 or Table 348.22 for metric designator 12 (trade size ⅜).

348.24 Bends—How Made

Bends must not damage the conduit and must not reduce the internal diameter of the conduit. Bends can be made with auxiliary equipment manually. The radius of the curve to the centerline can not be less than that shown in Table 344.24. Use the "Other Bends" column.

348.26 Bends—Number in One Run

The maximum number that is permitted is not more than 360 degrees in bends between pull points. The radius of curve of the inner edge of a field bend must comply with Table 346.10.

348.28 Trimming

348.30 Securing and Supporting

Securely fasten within 300 mm (12 in.) of a termination point and supports must be not more than 1.4 m (4½ ft.) apart. There is a provision for supports through framing openings. There are three exceptions.

348.42 Couplings and Connectors

348.56 Splices and Taps

They can be made only in accordance with 300.15.

350.60 Grounding and Bonding

An equipment grounding conductor must be installed when the conduit is used to connect equipment where flexibility is needed. The equipment grounding conductor must be installed in accordance with 250.134(B). Equipment bonding conductors must be installed in accordance with 250.102.

ARTICLE 350
LIQUIDTIGHT FLEXIBLE METAL CONDUIT
Type LFMC

I. General

350.1 Scope

350.2 Definition

This conduit is similar to flexible metal conduit, but it has a jacket over the metal conduit. The jacket is liquidtight, nonmetallic, and sunlight-resistant.

350.3 Listing Requirements

II. Installation

351.10 Uses Permitted

This conduit can be installed exposed or concealed, directly buried in earth and if protection from liquids, vapor, or solids is necessary or if flexibility is desired. It can be installed in hazardous areas if approved for that installation. Refer to Sections 501.4(B), 502.4, 503.3, 504.20, and 553.7(B).

350.12 Uses Not Permitted

It cannot be installed when subject to damage or where temperatures exceed that which is approved for the material.

350.20 Size

Minimum metric designator 16 (trade size ½). For exceptions, refer to Section 348.20(A). The maximum size allowed is metric designator 103 (trade size 4). Refer to 300.1(C) for metric designators and trade sizes.

350.22 Number of Conductors

Refer to Chapter 9, Table 1. For metric designator 12 (trade size ⅜) conduit, refer to Table 348.22 in the *NEC®*.

350.24 Bends—How Made

Bends must not damage the conduit and must not reduce the internal diameter of the conduit. Bends can be made with auxiliary equipment manually. The radius of the curve to the centerline can not be less than that shown in Table 344.24. Use the "Other Bends" column.

350.26 Bends—Number in One Run

The maximum that is permitted is not more than 360° in bends between pull points.

350.30 Securing and Supporting

It must be supported and secured not more than 1.4 m (4½ ft.) apart. It must be securely fastened within 300 mm (12 in.) of a conduit termination. Three exceptions are for when the conduit is fished, where flexibility is needed, length not exceeding 900 mm (3 ft.) at terminals, and lengths not exceeding 1.8 m (6 ft.) from a fixture terminal connection for tap conductors to a lighting fixture as noted in Section 410.67(C). Horizontal runs allowed to be supported by the openings in framing members not more than 1.4 m (4½ ft.) apart and fastened within 300 mm (12 in.) of a termination.

350.42 Couplings and Connectors

350.56 Splices and Taps

They can be made only in accordance with Section 300.15.

350.60 Grounding and Bonding

This type of conduit can be used as a means of grounding under certain conditions. Refer to the *NEC*®.

III. Construction Specifications

350.120 Marking

ARTICLE 352
RIGID NONMETALLIC CONDUIT
Type RNC

I. General

352.1 Scope

352.2 Definition

A conduit which is circular in cross section and made from a nonmetallic material.

352.6 Listing Requirements

The RNC, elbows, and associated fittings must be listed.

II. Installation

352.10 Uses Permitted

Some of the uses permitted and other references in the *NEC*® are: concealed in walls, floors, ceilings, and cinder fill; exposed when not subject to injury; in corrosive locations where approved for that type of location and in accordance with Section 300.6; in dry and damp locations in accordance with Section 352.12; in wet locations where the material is approved; and underground in accordance with Sections 300.5 and 300.50. There is a provision for use as a support in certain specific cases.

352.12 Uses Not Permitted

This type of conduit must not be used in hazardous locations except where permitted by Sections 503.3(A), 504.20, 514.8, 515.8, and the Class 1, Division 2 locations in the Exception to Section 501.4(B); in areas where there might be physical damage to the conduit; in theaters and similar locations except as noted in Articles 518 and 520; where the temperature rating of the conduit is exceeded by that of the conductor insulation; and in areas where the ambient temperature or the insulation temperature will be greater than allowed for the conduit and for supporting fixtures or equipment.

352.20 Size

The minimum size is metric designator 16 (trade size ½). The maximum size is metric designator 155 (trade size 6).

352.22 Number of Conductors

Refer to Chapter 9, Table 1.

352.24 Bends—How Made

The conduit must not be injured. Refer to Table 344.24.

352.26 Bends—Number in One Run

The maximum that is permitted is not more than 360° in bends between pull points.

352.28 Trimming

Remove rough edges on cuts.

352.30 Securing and Supporting

Refer to Section 300.18 and Table 352.30(B) for spacing of supports. It must be fastened no farther than 900 mm (3 ft.) from a termination point. If it is listed for other securing spacing, it is permitted to follow but not exceed the listing. There is an exception for supports by openings through framing members of horizontal runs.

352.44 Expansion Fittings

Expansion fittings have to be provided to compensate for thermal expansion and contraction where the change in length according to Table 352.44(A) or (B) will be 6 mm (¼ in.) or greater between securely mounted conduit terminations.

352.46 Bushings

A bushing must be used at the point where the conduit goes into a box or fitting unless protection is provided in the box or fitting.

352.48 Joints

Use approved methods.

352.56 Splices and Taps

They are permitted only in accordance with Section 300.15.

352.60 Grounding

A separate equipment grounding conductor must be installed. There are two exceptions.

III. Construction Specifications

352.100 Construction

RNC and fittings must be composed of a nonmetallic material. It must be resistant to moisture and chemical atmospheres. Above ground it must be flame resistant, resistant to impact and crushing, resistant to distortion from heat, and resistant to low temperature and sunlight. Below ground it must be resistant to corrosive agents and moisture, be able to withstand crushing and other abuse. If it is direct buried without encasement in concrete, it must be able to resist the continued loading it will encounter.

352.120 Marking

It must be marked at least every 3 m (10 ft.) in each length. There are requirements for marking on conduit used above and below ground.

ARTICLE 354
NONMETALLIC UNDERGROUND CONDUIT WITH CONDUCTORS
Type NUCC

I. General

354.1 Scope

354.2 Definition

This is a factory assembly of conductors or cables in a nonmetallic circular conduit.

TABLE 352.44(A) Expansion Characteristics of PVC Rigid Nonmetallic Conduit Coefficient of Thermal Expansion = 6.084×10^{-5} mm/mm/°C (3.38×10^{-5} in./in./°F)

Temperature Change (°C)	*Length Change of PVC Conduit (mm/m)*	*Temperature Change (°F)*	*Length Change of PVC Conduit (in./100 ft.)*	*Temperature Change (°F)*	*Length Change of PVC Conduit (in./100 ft.)*
5	0.30	5	0.20	105	4.26
10	0.61	10	0.41	110	4.46
15	0.91	15	0.61	115	4.66
20	1.22	20	0.81	120	4.87
25	1.52	25	1.01	125	5.07
30	1.83	30	1.22	130	5.27
35	2.13	35	1.42	135	5.48
40	2.43	40	1.62	140	5.68
45	2.74	45	1.83	145	5.88
50	3.04	50	2.03	150	6.08
55	3.35	55	2.23	155	6.29
60	3.65	60	2.43	160	6.49
65	3.95	65	2.64	165	6.69
70	4.26	70	2.84	170	6.90
75	4.56	75	3.04	175	7.10
80	4.87	80	3.24	180	7.30
85	5.17	85	3.45	185	7.50
90	5.48	90	3.65	190	7.71
95	5.78	95	3.85	195	7.91
100	6.08	100	4.06	200	8.11

TABLE 352.44(B) Expansion Characteristics of Reinforced Thermosetting Resin Conduit (RTRC)
Coefficient of Thermal Expansion = 2.7×10^{-5} mm/mm/°C (1.5×10^{-5} in./in./°F)

Temperature Change (°C)	*Length Change of PVC Conduit (mm/m)*	*Temperature Change (°F)*	*Length Change of PVC Conduit (in./100 ft.)*	*Temperature Change (°F)*	*Length Change of PVC Conduit (in./100 ft.)*
5	0.14	5	0.09	105	1.89
10	0.27	10	0.18	110	1.98
15	0.41	15	0.27	115	2.07
20	0.54	20	0.36	120	2.16
25	0.68	25	0.45	125	2.25
30	0.81	30	0.54	130	2.34
35	0.95	35	0.63	135	2.43
40	1.08	40	0.72	140	2.52
45	1.22	45	0.81	145	2.61
50	1.35	50	0.90	150	2.70
55	1.49	55	0.99	155	2.79
60	1.62	60	1.08	160	2.88
65	1.76	65	1.17	165	2.97
70	1.89	70	1.26	170	3.06
75	2.03	75	1.35	175	3.15
80	2.16	80	1.44	180	3.24
85	2.30	85	1.53	185	3.33
90	2.43	90	1.62	190	3.42
95	2.57	95	1.71	195	3.51
100	2.70	100	1.80	200	3.60

354.6 Listing Requirements

II. Installation

354.10 Uses Permitted

It can be used encased or embedded in concrete, in cinder fill, for direct burial cable with cover requirements as noted in Tables 300.5 and 300.50 under rigid nonmetallic conduit, and in underground locations subject to corrosive action as noted in Section 300.6 and where subject to chemicals when the assembly is approved for it.

354.12 Uses Not Permitted

It cannot be used in exposed locations, inside buildings with one exception, and in hazardous (classified) locations except as noted in Sections 503.3(A), 504.20, 514.8, 515.8, and 501.4(B).

354.20 Size

It cannot be smaller than metric designator 16 (trade size ½) or larger than metric designator 103 (trade size 4).

354.22 Number of Conductors

Not to exceed number shown in Table 1 of Chapter 9.

354.24 Bends—How Made

They must be made manually; the conduit must not be damaged and the internal diameter must not be reduced. Refer to Table 344.24.

354.26 Bends—Number in One Run

The maximum is not more than 360° between termination points.

354.28 Trimming

354.46 Bushings

354.48 Joints

354.50 Conductor Terminations

354.56 Splices and Taps

354.60 Grounding

III. Construction Specifications

354.100 Construction

354.120 Marking

Must be marked every 3.05 m (10 ft.). It also must be identified on a tag at each end of the assembly or on the side of the reel.

ARTICLE 356
LIQUIDTIGHT FLEXIBLE NONMETALLIC CONDUIT
Type LFNC

356.1 Scope

356.2 Definition

This conduit has a circular cross section. It comes in various types. It can have a smooth inner surface with integral reinforcement; smooth seamless inner core and cover bonded together with the reinforcement between the core and cover; or corrugated surfaces without reinforcement within the walls (Type LFNC-C). This type of conduit is flame-resistant and is approved for electrical conductors, with fittings.

356.6 Listing Requirements

II. Installation

356.10 Uses Permitted

This conduit can be installed exposed or concealed if protection from vapors, liquids, or solids is necessary, if flexibility is desired, or for outdoors if listed and marked for this purpose. It can also be used outdoors when listed and suitable and for direct burial use when listed and marked, as well as in lengths longer than 1.8 m (6 ft.), where additional length is essential. It can be used as a prewired assembly under certain conditions.

356.12 Uses Not Permitted

It can not be installed if subject to damage, when temperatures will exceed those for which the conduit is approved, in lengths in excess of 1.8 m (6 ft.) except under certain conditions and where voltages are higher than 600 V.

356.20 Size

Metric designator 16 (trade size ½) to metric designator 103 (trade size 4). For leads of motors metric designator 12 (⅜) can be used in accordance with 430.145(B). There are two other exceptions.

356.22 Number of Conductors

Refer to Chapter 9, Table 1.

356.24 Bends—How Made

They must be made manually; the conduit must not be damaged and the internal diameter must not be reduced. Refer to Table 344.24.

356.26 Bends—Number in One Run

The maximum that is permitted is not more than 360.

356.28 Trimming

356.30 Securing and Supporting

356.42 Couplings and Connectors

356.56 Splices and Taps

Can only be made in accordance with 300.15.

356.60 Grounding and Bonding

An equipment grounding conductor must be installed.

III. Construction Specifications

356.100 Construction

356.120 Marking

ARTICLE 358 ELECTRICAL METALLIC TUBING Type EMT

I. General

358.1 Scope

358.2 Definition

It is a thinwall listed metallic tubing, circular, with a cross-sectional area that is approved when joined together with listed fittings, for the installation of electrical conductors.

358.6 Listing Requirements

II. Installation

358.10 Uses Permitted

It can be used for both exposed and concealed work. Both ferrous and nonferrous tubing and fittings can be installed in concrete, in areas subject to severe corrosion, and in direct contact with the earth if protected against corrosion and suitable for the condition. It can be installed in wet locations if all supports, bolts, etc. are corrosion resistant or covered with corrosion resistant material.

358.12 Uses Not Permitted

It cannot be used where it will be subject to severe damage, where protected from corrosion by only enamel, in cinder concrete or cinder fill where there is permanent moisture unless there is a layer of noncinder concrete on all sides at least 50 mm (2 in.) thick or the tubing is at least 450 mm (18 in.) under the fill, to support fixtures or equipment except conduit bodies not larger than the largest trade size of the tubing, or in hazardous locations except as permitted in Sections 502.4, 503.3, and 504.20. Where possible, there should be no contact between dissimilar metals in the system.

358.20 Size

The minimum size permitted is metric designator 16 (trade size ½) unless used for enclosing motor leads [see Section 430.145(B)]. The maximum size permitted is metric designator 103 (trade size 4).

358.22 Number of Conductors

Refer to Chapter 9, Table 1.

358.24 Bends—How Made

The tubing must not be injured. Refer to Table 344.24 for the radius of the center line of a field bend.

358.26 Bends—Number in One Run

The maximum that is permitted is not more than 360° in bends between pull points.

358.28 Reaming and Threading

Threading is not permitted except on factory-threaded integral couplings. Remove rough edges of all cuts.

358.30 Securing and Supporting

The tubing must be installed as a complete system as noted in Article 300. The tubing must be fastened at intervals no longer than 3 m (10 ft.) and not more than 900 mm (3 ft.) from a box, fitting, etc. The 900 mm (3-ft.) rule may be extended to 1.5 m (5 ft.) for unbroken lengths where structural members interfere at 900 mm (3 ft.). There is an exception for concealed work. EMT in horizontal runs supported by openings through framing not more than 3 m (10 ft.) and fastened within 900 mm (3 ft.) of termination is permitted.

358.42 Couplings and Connectors

They must be tight and concrete-tight if used in concrete and raintight if used in wet locations.

358.56 Splices and Taps

They are to be made in accordance with Section 300.15.

358.60 Grounding

EMT is permitted as an equipment grounding conductor.

III. Construction Specifications

358.100 Construction

If an integral threaded coupling is used, threads for both the couplings and tubing must be factory made. The EMT threads and couplings can not bend the tubing at any part of the thread.

358.120 Marking

It must be clearly and durable marked at least every 3 m (10 ft.) on accordance with 110.21.

ARTICLE 360
FLEXIBLE METALLIC TUBING
Type FMT

I. General

360.1 Scope

360.2 Definition

It is flexible, metallic, circular in cross section, and liquidtight without a nonmetallic jacket.

360.6 Listing Requirements

II. Installation

360.10 Uses Permitted

This type of tubing can be used for branch circuits in dry locations, in accessible locations, concealed, and not more than nominal 1000-V systems.

360.12 Uses Not Permitted

It cannot be used where subject to physical damage, in hoistways, storage battery rooms, hazardous locations, embedded in concrete, or directly buried in the earth. Lengths larger than 1.8 m (6 ft.) are not permitted.

360.20 Size

The minimum size permitted is metric designator 16 (trade size ½) except that metric designator 12 (trade size ⅜) can be used in accordance with Section 300.22(B) and (C) or part of an approved lighting fixture assembly in accordance with Section 410.67(C) and not more than 1.8 m (6 ft.) in length. The maximum size permitted is metric designator 21 (trade size ¾).

360.22 Number of Conductors

Refer to Chapter 9, Table 1, for the number of conductors permitted in metric designator 16 (trade size ½) and metric designator 21 (trade size ¾) tubing. Refer to Table 348.22 for the number of conductors permitted in metric designator 12 (trade size ⅜) tubing.

360.24 Bends

If the tubing is bent infrequently after installation, refer to Table 360.24(A) in the *NEC®*. If the tubing is bent for installation, use Table 360.24(B) in the *NEC®*.

360.40 Boxes and Fittings

360.56 Splices and Taps

360.60 Grounding

It is permitted as an equipment grounding conductor only in accordance with 250.118(B)

III. Construction Specifications

360.120 Marking

Mark in accordance with 110.21.

ARTICLE 362
ELECTRICAL NONMETALLIC TUBING
Type ENT

I. General

362.1 Scope

362.2 Definitions

362.6 Listing Requirements

II. Installation

362.10 Uses Permitted

362.12 Uses Not Permitted

362.20 Size

362.22 Number of Conductors

362.24 Bends—How Made

362.26 Bends—Number in One Run

362.28 Trimming

362.30 Securing and Supporting

362.46 Bushings

362.48 Joints

362.56 Splices and Taps

362.60 Grounding

III. Construction Specifications

362.100 Construction

362.120 Marking

ARTICLE 366
AUXILIARY GUTTERS

366.1 Scope

366.2 Use

366.3 Extension Beyond Equipment

It must not extend more than 9 m (30 ft.) from the equipment it supplements.

366.4 Supports

366.5 Covers

366.6 Number of Conductors

Not more than 30 conductors are allowed at any point without the derating factor of 310.15(B)(2)(a) and the area of the conductors cannot be more than 20% of the area of a sheet metal auxiliary gutter. For signaling or motor controller circuits where the conductors are used only for starting duty, they are not considered current carrying conductors. The area of the conductors in a nonmettalic auxiliary gutter cannot be more than 20% of the area of the gutter.

366.7 Ampacity of Conductors

The following refer to sheet metal auxiliary gutters. When there are not more than 30 current-carrying conductors, the derating factors for ampacities do not have to be applied. The capacities of bars cannot exceed 1.55 A per square millimeter (1000 A per square inch) for copper and 1.09 A per mm^2 (700 A per square inch) for aluminum. For nonmetallic auxiliary gutters, use the derating factors in Section 310.15(B)(2)(a).

366.8 Clearance of Bare Live Parts

366.9 Splices and Taps

366.10 Construction and Installation

The following article titles are listed for reference:

Article 368—Busways; **Article 370**—Cablebuse; **Article 372**—Cellular Concrete Floor Raceways.

ARTICLE 374 CELLULAR METAL FLOOR RACEWAYS

374.1 Scope

374.2 Definitions

This is a raceway system that is part of the entire floor and installed as such. A complete definition of terms is given in the *NEC*®.

374.3 Uses Not Permitted

II. Installation

374.4 Size of Conductors

Not larger than 1/0 AWG unless special permission is obtained.

374.5 Maximum Number of Conductors in Raceway

Not more than 40% fill.

374.6 Splices and Taps

374.7 Discontinued Outlets

374.8 Markers

374.9 Junction Boxes

374.10 Inserts

374.11 Connection to Cabinets and Extensions from Cells

III. Construction Specifications

374.12 General

The following article titles are listed for reference:

Article 376—Metal Wireways; **Article 378**—Nonmetallic Wireways

ARTICLE 380
MULTIOUTLET ASSEMBLY

380.1 Scope

380.2 Use

380.3 Metal Multioutlet Assembly through Dry Partitions

ARTICLE 382
NONMETALLIC EXTENSIONS

I. General

382.1 Scope

382.2 Definition

This is a conductor arrangement with two insulated conductors in a nonmetallic jacket or thermoplastic cover.

II. Installation

382.10 Uses Permitted

Only from an existing 15- or 20-A outlet on a branch circuit. It must be run exposed and dry. For surface extensions the buildings must only be occupied for residential or office use and with height restrictions of three floors above grade.

382.12 Uses Not Permitted

This type of extension cannot be used in unfinished areas, where there are corrosive vapors, through a floor or partition; and where the voltage between conductors is greater than 150 V for surface work and 300 V for aerial cable.

382.15 Exposed

382.26 Bonds

382.30 Securing and Supporting

382.40 Boxes and Fittings

382.56 Splices and Taps

ARTICLE 384
STRUT-TYPE CHANNEL RACEWAY

I. General

384.1 Scope

384.2 Definition

384.6 Listing Requirements

II. Installation

384.10 Uses Permitted

384.11 Uses Not Permitted

384.21 Size of Conductors

384.22 Number of Conductors

384.30 Securing and Supporting

384.56 Splices and Taps

384.60 Grounding

III. Construction Specifications

384.100 Construction

384.120 Marking

ARTICLE 386
SURFACE METAL RACEWAYS

I. General

386.1 Scope

386.2 Definition

A metallic raceway with its associated fittings mounted to the surface of a structure.

386.6 Listing Requirements

II. Installation

386.10 Uses Permitted

It can be used in dry locations, in Class I Division 2 hazardous locations in accordance with 501.4(B) Exception, through walls and floors under certain conditions, and under raised floors in accordance with 656.5(D)(2).

386.12 Uses Not Permitted

It cannot be installed where it might be damaged, where there are corrosive vapors, in hoistways, where the voltage between conductors is 300 V or more unless the metal is at least 1.02 mm (0.04 in.), or concealed.

386.21 Size of Conductors

Conductors cannot be larger than the raceway is designed.

386.22 Number of Conductors or Cables

Refer to design criteria. Exceptions to the use of derating factors are noted in the *NEC*® in this paragraph.

386.56 Splices and Taps

386.60 Grounding

386.70 Combination Raceways

III. Construction Specifications

386.100 Construction

ARTICLE 388
SURFACE NONMETALLIC RACEWAYS

I. General

388.1 Scope

388.2 Definition

A nonmetallic raceway with its associated fittings mounted to the surface of a structure.

388.6 Listing Requirements

II. Installation

388.10 Uses Permitted

This can be used in dry locations. It can be used to pass transversely through dry walls, dry partitions, and dry floors if it is unbroken. You must maintain access to the conductors on either side of the wall floor or partition.

388.11 Uses Not Permitted

It cannot be installed concealed except as permitted in 388.10(2), where it might be damaged, where the voltage between conductors is 300 V or more unless so listed, in hoistways, in any hazardous area location except Class I, Division 2 permitted in 501.4(B), Exception, if ambient temperatures are higher than that for which the raceway is listed, or for conductors with insulation temperatures higher than the raceway is listed.

388.21 Size of Conductors

Not larger than the raceway design calls for.

388.22 Number of Conductors or Cables

Number not larger than the raceway design calls for.

388.56 Splices and Taps

388.60 Grounding

388.70 Combination Raceways

III. Construction Specifications

388.100 Construction

It is made of nonmetallic material that is resistant to moisture, chemical atmospheres, impact and crushing, distortion from heat, and low temperatures. It is also flame-retardant.

388.120 Marking

ARTICLE 390
UNDERFLOOR RACEWAYS

390.1 Scope

390.2 Use

(A) Permitted. This type of raceway can be used beneath the surface of concrete flooring or in an office flush with the concrete floor with other floor covering.

(B) Not Permitted. This type of raceway can not be used if exposed to corrosive vapors, in a hazardous location [for an exception refer to 501.4(B)(3) and 504.20], in concrete or in corrosive areas unless properly protected and approved.

390.3 Covering

390.4 Size of Conductors

Not larger than designed for.

390.5 Maximum Number of Conductors in Raceway

Not more than 40% fill.

390.6 Splices and Taps

390.7 Discontinued Outlets

390.8 Laid in Straight Lines

390.9 Marker at Ends

390.10 Dead Ends

390.13 Junction Boxes

390.14 Inserts

390.15 Connections to Cabinets and Walls

ARTICLE 392
CABLE TRAYS

392.1 Scope

392.2 Definition

392.3 Uses Permitted

Cable tray can be used as a support system for branch circuits, feeders, services, communication circuits, control circuits, and signaling circuits. It is not limited to industrial locations. If it is exposed to direct sunlight then jacketed cables or insulated conductors that are sunlight resistant must be used.

(A) Wiring Methods. Refer to the noted article or section for the conditions of use of the following:

	Article	Section
Armored cable	333	
Communication raceways	800	
Electrical metallic tubing	358	
Electrical nonmetallic tubing	362	
Fire alarm cables	760	
Flexible metal conduit	348	
Flexible metallic tubing	360	
Instrumentation tray cable	727	
Intermediate metal conduit	342	
Liquidtight flexible metal conduit	350	
Liquidtight flexible nonmetallic conduit	356	
Metal-clad cable	330	
Mineral-insulated, metal-sheathed cable	332	
Multiconductor service-entrance cable	338	
Multiconductor underground feeder and branch-circuit cable	340	
Multipurpose and communications cables	800	

	Article	Section
Nonmetallic-sheathed cable	334	
Power and control cable	336	
Power-limited tray cable		725.61(C) and 725.71(F)
Optical fiber cables	770	
Optical fiber racways	770	
Other factory-assembled, multiconductor control, signal, or power cables that are specifically approved for installation in cable trays		
Rigid metal conduit	344	
Rigid nonmetallic conduit	347	

(B) In industrial establishments. Wiring methods of 318.3(A) can be used in accordance with their respective article's requirements. Where only qualified maintenance personnel are available, certain cables are permitted in ladder, ventilated trough, and ventilated channel-type cable trays. Single-conductor cable must be 1/0 AWG or larger and be permitted to be used in cable tray. If the cable is sized 1/0 AWG through 4/0 AWG, it must be installed in ladder-type cable tray with a maximum rung spacing of 9 in. or in ventilated trough cable tray. Welding cable can be installed in accordance with Article 630, Part IV. Equipment grounding single conductors can be insulated, covered, or bare 4 AWG or larger. Multiconductor cable type MV can also be installed as noted above. Single conductors must be installed as per 318.3(B)(1).

(C) Equipment Grounding Conductors

(D) Hazardous (Classified) Locations

(E) Nonmetallic Cable Tray

392.4 Uses Not Permitted

Cable trays cannot be used where they are exposed to severe physical injury, in hoistways, and in environmental air-handling spaces. An exception is made in the latter case where used to support a wiring method noted in Section 300.22.

392.5 Construction Specifications

Cable trays must have proper strength and rigidity, smooth edges, be protected against corrosion, have side rails, and contain the proper fittings. Flame-retardant material must be used for nonmetallic cable tray.

392.6 Installation

The cable tray system must be installed completely as a system before the cables are installed, have the proper supports, and have covers where required. Multiconductor cables can be installed in the same tray if 600 V or less. Cables over 600 V cannot be installed with cables under 600 V, with two exceptions. The cable trays must be exposed and accessible and have room around them for adequate access. They can go through partitions, walls, and floors if installed in accordance with Section 300.21 (proper and approved firestopping methods used). Incidental support from cable tray is permitted where approved and designed to support the load. Additional requirements are noted.

392.7 Grounding

Proper grounding must be attained in compliance with Article 250. The entire system must be grounded. Minimum size steel and aluminum trays are indicated in Table 392.7(B) of the *NEC*®.

392.8 Cable Installation

(A) Splices are permitted if approved methods are used, they are accessible, and are not over the side rails.

(B) Cables must be fastened in sloping or vertical runs.

(C) If the cable is installed in a bushed conduit and tubing for protection or support, a box is not required.

(D) If installed in parallel, the conductors must be grouped with one conductor per phase in each group. Single conductors must be bound in circuit groups.

(E) Single conductors.

392.9 Number of Multiconductor Cables, Rated 2000 Volts, Nominal, or Less, in Cable Trays

Refer to this section in the Code and Table 392.9.

392.10 Number of Single Conductor Cables, Rated 2000 Volts or Less, in Cable Trays

Refer to this section in the Code and Table 392.10(A).

392.11 Ampacity of Cables Rated 2000 Volts or Less in Cable Trays

(A) Multiconductor Cables. Multiconductor cables, rated 2000 V or less, installed as per Section 392.9 are to follow ampacities in Tables 310.16 and 310.18. The derating factors of Section 310.15(B)(2)(a) apply only to multiconductor cables of three or more current carrying conductors in cable, not in cable tray. If the tray is covered for more than 6 ft. with a solid unventilated cover, the ampacities allowed are 95% of those shown in the tables. An additional exception is allowed for a single layer in uncovered trays.

(B) Single Conductor Cables. The derating factors of Section 310.15(B)(2)(a) don't apply to cables in cable tray. Single conductor or single conductors cabled together, rated 2000 V or less, properly installed in accordance with the Code, are to use ampacities as follows:

(1) Cables 600 kcmil and larger in uncovered trays use not more than 75% of the ampacities in Tables 310.17 and 310.19. If the tray is covered for more than 6 ft. with unventilated covers, use a maximum 70% of the ampacities in the tables.

(2) Cables 1/0 AWG through 500 kcmil in uncovered trays use not more than 65% of the ampacities in Tables 310.17 and 310.19. If the tray is covered for more than 6 ft. with unventilated covers, use a maximum 60% of the ampacities in the tables.

(3) If single conductor cables are placed in a single layer with at least one cable diameter separation, the tray is uncovered and the cable is 1/0 AWG or larger, the ampacities in Tables 310.17 and 310.19 may be used.

4. Special allowance when installed in uncovered tray with a triangular configuration.

392.12 Number of Type MV and Type MC Cables (2001 Volts or Over) in Cable Trays

392.13 Ampacity of Type MV and Type MC Cables (2001 Volts or Over) in Cable Trays

ARTICLE 394
CONCEALED KNOB-AND-TUBE WIRING

I. General

394.1 Scope

394.2 Definition

This type of wiring is a single insulated conductor using knobs, tubes, and flexible nonmetallic tubing for support and protection.

II. Installation

394.10 Uses Permitted

It can be used only to extend an existing installation or elsewhere with special permission from the authorities having jurisdiction and in hollow spaces of walls and ceilings and unfinished attic and roof spaces (refer to Section 394.23 in the *NEC*®).

394.12 Uses Not Permitted

It is not allowed in commercial garages, motion picture studios, classified hazardous locations, theaters and similar locations, or in hollow spaces of walls, ceilings, and attics under certain conditions.

394.17 Through or Parallel to Framing Members

394.19 Clearances

394.23 In Accessible Attics

394.30 Securing and Supporting

394.42 Devices

394.56 Spplices and Taps

III. Construction Specifications

394.104 Conductors

ARTICLE 396
MESSENGER SUPPORTED WIRING

I. General

396.1 Scope

396.2 Definition

This type of wiring is exposed using a messenger wire as a support for insulated conductors. These conductors are usually in the form of a type of cable. In some instances in industrial establishments, insulated wire can be used as well as cable.

II. Installation

396.10 Uses Permitted

The following uses are permitted under the noted requirements:

(A) Cable Types

	Section	Article
Metal-clad cable		330
Mineral-insulated, metal-sheathed cable		332
Multiconductor service-entrance cable		338
Multiconductor underground feeder and branch-circuit cable		340
Other factory-assembled, multiconductor control, signal, or power cables that are identified for the use		
Power and control tray cable		336
Power-limited tray cable	725.61(C) and 726.71(E)	

(B) In Industrial Establishments.

(C) Hazardous (Classified) Location

396.12 Uses Not Permitted

396.30 Messenger Support

396.56 Conductor Splices and Taps

396.60 Grounding

ARTICLE 398
OPEN WIRING ON INSULATORS

I. General

398.1 Scope

398.2 Definition

This type of wiring is exposed with a single insulated conductor using an insulator (knob, cleat, tube, or flexible tubing) for protection and support.

II. Installation

398.10 Uses Permitted

This type of wiring can be used only where the nominal voltage does not exceed 600 V, in industrial or agricultural structures, both indoors and outdoors, both wet and dry locations, for services, and where corrosive vapors are present.

398.12 Uses Not Permitted

Cannot be used if concealed by the building structure.

398.15 Exposed Work

398.17 Through or Parallel to Framing Member

398.19 Clearances

398.23 In Accessible Attics

398.30 Securing and Supporting

398.42 Devices

III. Construction Specifications

398.104 Conductors

4

EQUIPMENT FOR GENERAL USE

ARTICLE 400
FLEXIBLE CORDS AND CABLES

I. General

400.1 Scope

400.2 Other Articles

400.3 Suitability

400.4 Types

Table 400.4 lists the approved types of flexible cables and cords. A cable or cord not on this list must have a special investigation. Notes to the table should be reviewed.

400.5 Ampacity of Flexible Cords and Cables

Tables 405.5(A) and 405.5(B) show the ampacity of flexible cords or cables of up to three current-carrying conductors. When more than three current-carrying conductors are used, a derating factor is used. These derating factors are shown in a table in the code. There is information on Ultimate Insulation Temperature. There are additional requirements.

400.6 Marking

400.7 Uses Permitted

(1) Pendants; (2) wiring of luminaires (fixtures); (3) connection of portable lamps or appliances; (4) for elevator cables; (5) cranes and hoists; (6) for utilization equipment if they are to be moved a lot; (7) to prevent noise and vibration transmission; (8) where removal for maintenance is required and the appliance is to use a flexible cord or cable; (9) for data processing cables (refer to Section 645.5); (10) connection of moving parts; (11) as allowed in Sections 527.4(B) and 527.4(C) for temporary wiring. When used as indicated in (3), (6), and (8), a plug must be used on the cord and it must be plugged into a receptacle for power.

400.8 Uses Not Permitted

Unless permitted in Section 400.7, they are not permitted: (1) for a substitute for fixed wiring of a building; (2) installed through a hole in walls, ceilings, or floors; (3) through doorways, windows, or similar openings; (4) installed concealed; (5) attached to building surfaces; (6) installed in raceways unless permitted by the code. There is an exception in accordance with Section 364.8.

400.9 Splices

400.10 Pull at Joints and Terminals

Some type of strain relief must be provided (i.e., a knot, tape, or special fitting).

400.11 In Show Windows and Show Cases

Use only Type S, SO, SE, SEO, SEOO, SOO, SJ, SJE, SJO, SJEO, SJEOO, SJOO, ST, STO, STOO, SJT, SJTO, SJTOO, SEW, SEOW, SEOOW, SJEW, SJEOW, SJEOOW, SJOW, SJTOW, SJTW, SJOOW, SOW, SOOW, SJTOOW, STW, STOW, or STOOW. This restriction is not necessary for chain-supported lighting fixtures or portable lamps and display merchandise.

400.12 Minimum Size

Refer to Table 400.4.

400.13 Overcurrent Protection

If not smaller than 18 AWG or tinsel cords or a cord approved for a specific appliance, no separate overcurrent protection is required. They are protected in accordance with 240.4.

400.14 Protection from Damage

II. Construction Specifications

400.20 Labels

400.21 Nominal Insulation Thickness

Refer to Table 400.4.

400.22 Grounded-Conductor Identification

One conductor that is used as a grounded circuit conductor must be clearly marked. Identification methods include (1) a white or gray braid, (2) a tracer of a distinguishable color in the braid, (3) white or gray insulation, (4) white or gray separator, (5) tinned conductor, and (6) some distinguishable surface marking.

400.23 Equipment Grounding-Conductor Identification

This conductor must also have a distinguishing marking of a colored braid, colored insulation, or colored cover. The color to be used is either green or green with yellow stripes.

400.24 Attachment Plugs

III. Portable Cables Over 600 Volts, Nominal

400.30 Scope

400.31 Construction

400.32 Shielding

400.33 Grounding

400.34 Minimum Bending Radii

400.35 Fittings

400.36 Splices and Terminations

ARTICLE 402
FIXTURE WIRES

402.1 Scope

402.2 Other Articles

402.3 Types

Refer to Table 402.3 for those types permitted. They can be used at 600 Volts unless otherwise specified.

402.5 Allowable Ampacities for Fixture Wires

Table 402.5 specifies the ampacity of fixture wire.

402.6 Minimum Size

Not smaller than 18 AWG.

402.7 Number of Conductors in Conduit or Tubing

Refer to Chapter 9, Table 1.

402.8 Grounded-Conductor Identification

402.9 Marking

402.10 Uses Permitted

Luminaires (lighting fixtures) and similar equipment where enclosed and protected and for connecting luminaires (lighting fixtures) to branch circuits.

402.11 Uses Not Permitted

Cannot be used as branch-circuit conductors.

402.12 Overcurrent Protection

As specified in 240.5.

ARTICLE 404
SWITCHES

I. Installation

404.1 Scope

404.2 Switch Connections

Three- and four-way switches must be installed so that switching is in the ungroundedconductor. If the wiring is installed in metal raceways or metal-armored cable, it must be in accordance with the requirements of Section 300.20(A). A grounded conductor must not be switched. There are two exceptions.

404.3 Enclosure

404.4 Wet Locations

404.5 Time Switches, Flashers, and Similar Devices

405.6 Position and Connection of Knife Switches

404.7 Indicating

404.8 Accessibility and Grouping

404.9 Provision for General-Use Snap Switches

404.10 Mounting of Snap Switches

404.11 Circuit Breakers at Switches

404.12 Grounding of Enclosures

404.13 Knife Switches

404.14 Rating and Use of Snap Switches

II. Construction Specifications

404.15 Marking

They must be marked with current, voltage, and the maximum horsepower if horsepower rated.

404.16 600-Volt Knife Switches

404.17 Fused Switches

Parallel fusing is not permitted except as noted in Section 240.8.

404.18 Wire Bending Space

ARTICLE 406
RECEPTACLES, CORD CONNECTORS, AND ATTACHMENT PLUGS (CAPS)

406.1 Scope

406.2 Receptacle Rating and Type

(A) Receptacles. They must be listed for the purpose. They must also be marked with the voltage and ampere ratings and well as the manufacturer's name or identification.

(B) Rating. They can not be rated for less than 15 amperes, 125 volts or 15 amperes, 250 volts. They can not be suitable for lampholders.

(C) Receptacles for Aluminum Conductors. If rated at 20 amperes or less and designed for use with aluminum conductors, they must by marked CO/ALR.

(D) Isolated Ground Receptacles. If there is an isolated ground connection for the reduction of electrical noise it must have an orange triangle on the face for identification. There is additional information noted on these types of receptacles.

406.3 General Installation Requirements

They must be located in branch circuits in accordance with Part III of Article 210.

(A) Grounding Type. If installed on a 15 or 20 ampere branch circuit the receptacle must be of the grounding type. They must be installed on circuits of the voltage and current for which they are rated. There are some exceptions.

(B) To Be Grounded. If the receptacle and cord connector has a grounding contact the contact must be effectively grounded. There are 2 exceptions.

(C) Methods of Grounding. The grounding contacts must be grounded by connection to the equipment grounding conductor of the supply circuit. The branch circuit has to have an equipment grounding conductor.

(D) Replacements. If a grounding means exists, then the replacement must be a grounding type receptacle. If a ground fault circuit interrupter protected receptacle is required by the code then it must be used as the replacement. The reader is referred to the code for three requirements for replacement of a nongrounding type receptacle.

(E) Cord and Plug Connected Equipment.

(F) Noninterchangeable Types.

406.4 Receptacle Mounting

There are six requirements listed in the *Code®*. Of note is the requirement that after installation the receptacle must be flush with or project from insulating material and must project a minimum of 0.4 mm (0.015 in.) from a metal faceplate.

406.5 Receptacle Faceplates (Cover Plates)

406.6 Attachment Plugs

406.7 Noninterchangeability

406.8 Receptacles in Damp or Wet Locations

Receptacles located in damp areas must have a weather proof enclosure. The receptacle need only be protected, not when it is in use. Receptacles located in wet location must be weatherproof protected at all times, if not attended. If it is attended while in use it has to be weatherproof only when the plug is removed. Receptacles located outdoors must be installed high enough so that water will not touch the plate. Floor receptacles must be installed so as not to be damaged by cleaning equipment. If the receptacle is flush mounted on an outdoor box, the faceplate assembly must provide the watertight connection. A receptacle can not be installed within a bathtub or shower space.

406.9 Grounding-Type Receptacles, Adapters, Cord Connectors, and Attachment Plugs.

Grounding-type receptacles must have a grounding slot. Grounding-type plugs must have a grounding prong and this prong must make contact first. The ground pole must be distinctively marked and it must not be used for any other purpose. There are requirements listed for grounding-pole identification. The reader is referred to the *Code*® for all the requirements.

406.10 Connecting Receptacle Grounding Terminal to Box

An equipment bonding jumper is to be used. There are four other methods that can be used for specific conditions.

ARTICLE 408
SWITCHBOARDS AND PANELBOARDS

I. General

408.1 Scope

Switchboards, panelboards, distribution boards, and battery-charging panels are covered.

408.2 Other Articles

Articles 240, 250, 314, 312, and 404 and applicable sections of other Code articles must be complied with.

408.3 Support and Arrangement of Busbars and Conductors

They must be located to be free from damage and be secured. They must be arranged to avoid overheating due to induction. When used as service equipment, a grounding bar must be provided and it must be bonded to the case and frame. There is an exception for high-impedance grounded neutral system connections. Locations of terminals must be so as to avoid the necessity of reaching across ungrounded line bus. In a four-wire delta system the bar or conductor with the highest voltage to ground must be marked. Three-phase buses must have the phases A, B, C, front to back, left to right, or top to bottom when looking from the front. There is an exception for equipment in the same switchboard or panelboard as the meter. Refer to Section 373.6 for minimum wire bending space. Except for interconnections ands control wiring, only conductors intended for termination in a vertical section of a switchboard can be located there. There is an exception to this.

408.4 Circuit Directory

All circuits and modifications must be noted on a directory on the face or inside of the panelboard or on a switch on a switchboard.

II. Switchboards

408.5 Location of Switchboards

408.6 Switchboards in Damp or Wet Locations

408.7 Location Relative to Easily Ignitable Material

408.8 Clearances

At least 900 mm (3 ft.) to a combustible ceiling. Around switchboard comply with 110.26.

408.9 Conductor Insulation

408.10 Clearance for Conductors Entering Bus Enclosure

408.12 Grounding of Instruments, Relays, Meters, and Instrument Transformers on Switchboards

Refer to Sections 250.17 through 250.178.

III. Panelboards

408.13 General

They must have current, voltage, phase, and manufacturer's trademark markings visible after installation. The rating may not be less than the feeder capacity. All circuits and modifications must be noted on a directory on the panelboard.

408.14 Classification of Panelboards

(A) Lighting and Appliance Branch-Circuit Panelboard. It falls into this category if 10% or more of the overcurrent devices protect lighting and appliance branch circuits which are circuits that are 30 amperes or less and have a neutral connection.

(B) Power Panelboard. It falls into this category if there are less than 10% of the overcurrent devices which protect lightning and appliance branch circuits.

408.15 Number of Overcurrent Devices on One Panelboard

Not more than 42 overcurrent devices can be installed in a lighting or appliance branch-circuit panelboard. This does not include mains. Each pole of a multipole circuit breaker is considered a single circuit.

408.16 Overcurrent Protection

(A) Individual overcurrent protection must be provided for a lighting and appliance branch-circuit panelboard of not more than two sets of circuit breakers or fuses. The overcurrent protection cannot exceed the ratings of the panelboard. This is not required if the feeder has overcurrent protection that does not exceed the rating of the panelboard. If an existing panelboard is used as service equipment in a single-residence occupancy, the individual protection is not required.

(B) A power panelboard with a supply neutral and more than 10% of the overcurrent devices protecting branch circuits 30 amperes or less must have supply side protection not greater than the rating of the panelboard.

(C) Overcurrent protection not greater than 200 A is required for a panelboard that has snap switches rated 30 A or less.

(D) For panelboards that are supplied through a transformer, the overcurrent protection required must be on the secondary of the transformer. There is an exception.

(E) A panelboard with less than three-phase buses cannot use a three-phase overcurrent device or disconnect. Delta breakers cannot be installed in panelboards.

(F) An additional means of fastening that requires a release other than a pull to release the device must be used for backfeed plug-in devices (plug-in overcurrent or plug-in main lug assemblies).

408.17 Panelboards in Damp or Wet Locations

Refer to Section 312.2(A).

408.18 Enclosure

Panelboards must be in some type of enclosure with a dead front. If only qualified persons have access to the panelboards, dead fronts are not required.

408.19 Relative Arrangement of Switches and Fuses

Fuses must be installed on the load side of switches. There is an exception for service equipment.

408.20 Grounding of Panelboards

Metal cabinets and frames must be in contact with each other and grounded. A terminal bar for grounding conductors must be inside the cabinet when separate grounding conductors or nonmetallic raceway or cable is used. If the panelboard frame and cabinet are metal, the terminal bar is bonded to them. If nonmetallic, the terminal bar is connected to the grounding conductor from the feeder. The terminal bar used for the grounding conductors must be identified and located where the connection is made from the grounded conductor to a grounding electrode in accordance with Article 250. An isolated ground conductor can pass through the panelboard with connection to the ground bar in accordance with Section 250.146(D).

408.21 Grounded Conductor Terminations

IV. Construction Specifications

408.30 Panels

Must be made of moisture-resistant noncombustible material for switchboards.

408.31 Busbars

Must be rigidly mounted.

408.32 Protection of Instrument Circuits

Instruments must be protected and the rating must not exceed 15 A.

408.33 Component Parts

They must be in accordance with Articles 240 and 380.

408.35 Wire Bending Space in Panelboards

Top and bottom wire bending space must be in accordance with Table 312.6(B). Side bending space must be in accordance with Table 312.6(A). Refer to this article in the *NEC*® for exceptions.

408.36 Minimum Spacings

Refer to Table 408.36.

ARTICLE 410
LUMINAIRES (LIGHTING FIXTURES), LAMPHOLDERS, AND LAMPS

I. General

410.1 Scope

410.2 Application of Other Articles

410.3 Live Parts

II. Luminaire (Fixture) Locations

410.4 Luminaires (Fixtures) in Specific Locations

(A) Wet and Damp Locations. The fixtures must be installed in a manner so that water will not enter or collect in the lampholder, wireway, etc.

(B) Corrosive Locations

(C) In Ducts or Hoods. This location is permitted only in nonresidential cooking areas and is subject to four specific conditions.

(D) Bathtub and Shower Areas

410.5 Luminaires (Fixtures) Near Combustible Material

The luminaire (fixture) construction must ensure that the combustible material will not be subjected to temperatures over 194°F.

410.6 Luminaires (Fixtures) over Combustible Material

These must be unswitched lampholders. The lampholder must be at least 2.5 m (8 ft.) above the floor unless an individual switch is installed for each fixture.

410.7 Luminaires (Fixtures) in Show Windows

External wiring is not permitted except chain-supported luminaires (fixtures).

410.8 Luminaires (Fixtures) in Clothes Closets

(A) Definition. The code has a specific set of dimensions for storage space with an explanatory diagram.

(B) Luminaire (Fixture) Types Permitted. Surface-mounted or recessed incandescent or fluorescent fixtures are permitted. The incandescent fixture must have a completely enclosed lamp.

(C) Luminaire (Fixture) Types Not Permitted. Pendant luminaires (fixtures) or lampholders and incandescent luminaires (fixtures) whose lamps are not totally enclosed.

(D) Location. Surface-mounted luminaires (fixtures) can be installed on the wall above the door or on the ceiling. The clearance between the luminaire (fixture) and nearest point of storage must be 300 mm (12 in.) for an incandescent luminaire (fixture) and 150 mm (6 in.) for a fluorescent luminaire (fixture). Recessed luminaires (fixtures) can be installed in the wall or ceiling. The clearance between the luminaire (fixture) and nearest point of storage must be 150 mm (6 in.).

410.9 Space for Cove Lighting

III. Provisions at Luminaire (Fixture) Outlet Boxes, Canopies, and Pans

410.10 Space for Conductors

410.11 Temperature Limit of Conductors in Outlet Boxes

410.12 Outlet Boxes to Be Covered

410.13 Covering of Combustible Material at Outlet Boxes

410.14 Connection of Electric-Discharge Luminaires (Lighting Fixtures)

IV. Luminaire (Fixture) Supports

410.15 Supports

410.16 Means of Support

The *NEC*® specifies a few requirements for outlet boxes, inspection, suspended ceilings, luminaire (fixture) studs, insulating joints, raceway fittings, busways, and trees. One important requirement to note is that a fixture that weighs more than 22.7 kg (50 lb.) cannot be supported from an outlet box. It must be independently supported.

V. Grounding

410.17 General

410.18 Exposed Luminaire (Fixture) Parts

If the wiring system provides a means of grounding, exposed conductive parts must be grounded. If no provision for grounding is available, they must be insulated. There is an exception for replacement lumianires (fixtures).

410.20 Equipment Grounding Conductor Attachment

A means to connect an equipment ground must be provided for a metal luminaire (fixture).

410.21 Methods of Grounding

Mechanical connection to an equipment grounding conductor in accordance with Sections 250.118 and 250.122.

VI. Luminaires (Wiring) of Fixtures

410.22 Luminaire (Fixture) Wiring—General

410.23 Polarization of Luminaires (Fixtures)

410.24 Conductor Insulation

410.27 Pendant Conductors for Incandescent Filament Lamps

410.28 Protection of Conductors and Insulation

410.29 Cord-Connected Showcases

410.30 Cord-Connected Lampholders and Luminaires (Fixtures)

410.31 Luminaires (Fixtures) as Raceways

Normally, luminaires (fixtures) cannot be used as raceways. It is permitted, however, if the luminaire (fixture) is listed for that use.

410.32 Wiring Supplying Luminaires (Fixtures) Connected Together

410.33 Branch-Circuit Conductors and Ballasts

VII. Construction of Luminaires (Fixtures)

410.34 Combustible Shades and Enclosures

410.35 Luminaire (Fixture) Rating

410.36 Design and Material

410.37 Nonmetallic Luminaires (Fixtures)

410.38 Mechanical Strength

410.39 Wiring Space

410.42 Portable Lamps

410.44 Cord Bushings

410.45 Tests

410.46 Live Parts

VIII. Installation of Lampholders

410.47 Screw-Shell Type

410.48 Double-Pole Switched Lampholders

410.49 Lampholders in Wet or Damp Locations

IX. Construction of Lampholders

410.50 Insulation

410.52 Switched Lampholders

X. Lamps and Auxiliary Equipment

410.53 Bases, Incandescent Lamps

410.54 Electric-Discharge Lamp Auxiliary Equipment

XI. Special Provisions for Flush and Recessed Luminaires (Fixtures)

410.64 General

This type of luminaire (fixture) must comply with Sections 410.65 through 410-72.

410.65 Temperature

They must be installed so that any adjacent combustible material will not be subjected to a temperature higher than 90°C (194°F). If the luminaire (fixture) is recessed in fire-resistant construction in a fire-resistant building, the temperature restriction is increased to 150°C (302°F) if the luminaire (fixture) is approved for this service. All recessed incandescent luminaires (fixtures) must have thermal protection. This restriction is not required if the luminaire (fixture) is installed in poured concrete and is approved for this use or when the luminaire (fixture) by construction design provides characterisitics that are equivalent to thermally protected characteristics, and are identified as such.

410.66 Clearance and Installation

Recessed parts of a luminaire (fixture) that is not identified for contact with combustible materials must be at least 13 mm (½ in.) from combustible material and at least 75 mm (3 in.) from thermal insulation above unless the luminaire (fixture) is approved for direct contact with the insulation.

410.67 Wiring

The insulation used on conductors must be suitable for the temperatures that they will be subjected to. Tap conductors can run to a box at least 300 mm (1 ft.) from the fixture. The raceway used for the tap conductor must be at least 450 mm (18 in.) long and not more than 1.8 m (6 ft.) long, be a suitable raceway or the tap conductor must be Type AC or MC cable.

XII. Construction of Flush and Recessed Luminaires (Fixtures)

410.68 Temperature

410.70 Lamp Wattage Marking

410.71 Solder Prohibited

410.72 Lampholders

XIII. Special Provisions for Electric-Discharge Lighting Systems of 1000 Volts or Less

XIV. Special Provisions for Electric-Discharge Lighting Systems of More Than 1000 Volts

XV. Lighting Track

ARTICLE 411
LIGHTING SYSTEMS OPERATING AT 30 VOLTS OR LESS

411.3 Listing Required

411.4 Locations Not Permitted

411.5 Secondary Circuits

411.6 Branch Circuit

411.7 Hazardous (Classified) Locations

ARTICLE 422
APPLIANCES

I. General

422.1 Scope

This covers electric appliances used in any type of occupancy.

422.3 Other Articles

All other parts of the Code apply when applicable. When used in hazardous areas, Articles 500 through 517 apply. Article 430 applies to motor appliances and Article 440 applies to appliances with hermetic refrigerant motors.

422.4 Live Parts

No live parts can be normally exposed other than open resistance heating elements such as those in toasters.

II. Installation

422.10 Branch-Circuit Rating

Generally, the rating of the circuit must be at least that of the total rating of the appliance for an individual appliance branch circuit and in accordance with Section 422.62. Motor circuits not having a marked rating must be in accordance with Part II of Article 430. If the appliance is to be continuously loaded (not a motor), the branch circuit must be rated at least 125% of the appliance unless the branch-circuit device and assembly are rated for continuous loading at 100%, in which case the branch circuit cannot be less than 100% of the appliance. Branch circuits for household cooking appliances are to be rated as provided in Table 220.19. If the branch circuit supplies an appliance and other loads, use the provisions of Section 210.23.

422.11 Overcurrent Protection

Follow (a) through (g) below and Section 422,10

(A) Branch-Circuit Overcurrent Protection. Must be protected in accordance with Section 240.4. If a protective device rating is marked on an appliance, this rating must not be exceeded.

(B) Household-Type Appliance with Surface Heating Elements. If its demand is more than 60 A when calculated with Table 220.19, it must have its power supply subdivided with no circuit more than 50 A.

(C) Infrared Lamp Commercial and Industrial Heating Appliances. Cannot be greater than 50 A.

(D) Open-Coil or Exposed Sheathed-Coil Types of Surface Heating Elements in Commercial-Type Heating Appliances. Cannot be greater than 50 A.

(E) Single Nonmotor-Operated Appliance. The branch-circuit overcurrent device rating should not exceed the protective device rating noted on the unit. If no rating is ntoed, use 20 A or less if the unit rating is not more than 13.3 A and 150% or less of the unit rating if it is more than 13.3 A.

(F) Electric Heating Appliances Employing Resistance-Type Heating Elements Rated More than 48 Amperes. Factory-installed overcurrent devices which must be used in this case are discussed in this paragraph.

(G) Motor-Operated Appliances. Overload protection for motor-operated appliances has to be as noted in Part III of Article 430. Hermetic refrigerant compressors must be as noted in Part VI of Article 440. When the appliance overcurrent devices required are separate from the appliance, then the device must be marked with information that can be used to make the selection with a minimum rating that can be used as specified in Sections 430.7 and 440.4.

422.12 Central Heating Equipment

Must be connected to an individual branch circuit except for fixed electric space heating equipment.

422.13 Storage-Type Water Heaters

422.14 Infrared Lamp Industrial Heating Appliances

422.15 Central Vacuum Outlet Assemblies

422.16 Flexible Cords

Specific types of cords are indicated for specific uses, with noted exceptions.

422.17 Protection of Combustible Material

422.18 Support of Ceiling-Suspended (Paddle) Fans

If the fan and its accessories do not weigh more than 16 kg (35 lb.), then it can be suspended from the outlet box. If the weight with the accessories is more than 16 kg (35 lb.), then it has to be suspended independently.

422.20 Other Installation Methods

III. Disconnecting Means

422.30 General

There must be a disconnecting means to disconnect the ungrounded conductors from the appliance. If there is more than one source to the appliance, then the disconnecting means have to be grouped and identified.

422.31 Disconnection of Permanently Connected Appliances

If the device is not over 300 VA or ⅛ hp, the branch-circuit overcurrent device can be the disconnecting means. If the ratings are higher, the branch-circuit overcurrent device can be used as the disconnecting means only if it can be locked in the open position and is in sight of the user. Refer to Section 422.33 for unit-switched appliances.

422.32 Disconnecting Means for Motor-Driven Appliances

In the case of a permanently installed motor-driven appliance that is larger than ⅛ hp, a switch or circuit breaker can be used if it is in sight of the appliance. It must then comply with Part X of Article 430.

422.33 Disconnection of Cord- and Plug-Connected Appliances

For these types of appliances an accessible separable connector or plug and receptacle can be used as the disconnecting means. Where it is not accessible a disconnecting means in accordance with Section 422.31 must be used. In the case of the household range, the plug and receptacle must be accessible from the front either directly or through removal of a drawer. The receptacle or connector must be rated at least the same as the appliances.

Demand factors are permitted if authorized in other sections of the *NEC*®.

422.34 Unit Switch(es) as Disconnecting Means

A unit switch or switches that is part of the appliance can be considered the disconnecting means for the appliance if it has a position marked "off," disconnects all ungrounded conductors and other disconnecting means are installed as follows:

(A) In multifamily dwellings this other disconnecting means must be installed at least on the same floor that the appliance is installed. It can control other appliances and lamps.

(B) In two-family dwellings this other disconnecting means can be installed outside the dwelling unit and it can be a switch for the dwelling unit, such as in a panel in the basement.

(C) The service disconnect will fulfill this other requirement in a one-family dwelling.

(D) In all other occupancies, if readily accessible, the branch unit switch or circuit breaker can be used as the other disconnecting means.

422.35 Switch and Circuit Breaker to Be Indicating

IV. Construction

422.40 Polarity in Cord- and Plug-Connected Appliances

The attachment plug must be polarized or of the grounding type if the appliance has a single-pole line connected manually operated switch; an Edison-base lampholder; or a 15- or 20-A receptacle. Polarity of Edison-base lampholders are noted in Section 410.42(A). There is an exception for a hand-held double-insulated shaver.

422.41 Cord- and Plug-Connected Appliances Subject to Immersion

A portable free-standing hydromassage unit or hand-held hair dryer has to be made to protect the user from electrocution if the unit becomes immersed when in the on or off position.

422.42 Signals for Heated Appliances

422.43 Flexible Cords

Specific types of cords are indicated for specific uses, with noted exceptions.

422.44 Cord- and Plug-Connected Immersion Heaters

422.45 Stands for Cord- and Plug-Connected Appliances

422.46 Flatirons

422.47 Water Heater Controls

422.48 Infrared Lamp Industrial Heating Appliances

422.49 High-Pressure Spray Washers

422.50 Cord- and Plug-Connected Heating Assemblies

V. Marking of Appliances

422.60 Nameplate

An easily accessible or visible nameplate with a name and rating in volts and amperes or volts and watts must be installed. Frequency must be noted when applicable.

422.61 Marking of Heating Elements

They must be marked if rated over 1 A, are replaceable in the field, and are part of an appliance.

422.62 Appliances Consisting of Motors and Other Loads

ARTICLE 424
FIXED ELECTRIC SPACE HEATING EQUIPMENT

I. General

424.1 Scope

This article does not apply to heating and room air conditioning. It does apply to heating cables, unit boilers, central systems, and other fixed electric space heating equipment.

424.2 Other Articles

All applicable rules of the *NEC*® apply.

424.3 Branch Circuits

(A) Branch-Circuit Requirements. It can be any size as long as it is an individual branch circuit supplying only one appliance. If more than one appliance is on the circuit, 30 A is the largest rating permitted (50 A for nonresidential infrared heating equipment).

(B) Branch-Circuit Sizing. Generally, the branch-circuit size and overcurrent rating must be at least 125% of the rating of the appliance. There are a few exceptions. These include contactors and relays [refer to Section 210.19(A), Exception], and fixed electric space heating equipment, including hermetic refrigeration equipment (refer to Sections 440.34 and 440.35).

II. Installation

424.9 General

424.10 Special Permission

424.11 Supply Conductors

424.12 Locations

This equipment cannot be installed where it may be severely damaged unless protected. When installed in damp or wet locations, the equipment has to be approved for such locations.

424.13 Spacing from Combustible Materials

Adequate spacing must be provided unless approved for direct contact.

III. Control and Protection of Fixed Electric Space Heating Equipment

424.19 Disconnecting Means

This must be provided. When more than one source is used, all disconnecting means must be grouped and marked.

(A) Heating Equipment with Supplementary Overcurrent Protection. Must be in sight of the supplementary overcurrent device and on the supply side of the overcurrent device if fuses and in addition item 1 or 2 following.

1. *Heater containing no motor rated over ⅛ horsepower.* This disconnecting means or a unit switch complying with Section 424.19(C) can be used as the disconnecting means for the motor controller and heater if it is in sight of the motor controller and heater or can be locked in the open position.

2. *Heater containing a motor(s) rated over ⅛ horsepower.* The same holds true as noted in the preceding subparagraph (1). In addition, if it is not in sight of the heater, a separate disconnect complying with Section 424.19(C) can be installed. If it is not in sight of the motor controller location, a disconnect means must be installed in accordance with Section 430.102, and where the motor is not in sight of the controller must comply with Section 430.102(B).

(B) Heating Equipment without Supplementary Overcurrent Protection

1. *Without motor or with motor not over ⅛ horsepower.* The branch-circuit switch or circuit breaker can serve as the disconnecting means if the switch or circuit breaker can be locked in the open position or is within sight of the heater.
2. *Over ⅛ horsepower.* A disconnecting means must be located within sight of the motor controller except when installed in accordance with Section 424.19(A)(2).

(C) Unit Switches as Disconnecting Means. A unit switch that is part of the heater disconnects all ungrounded conductors and has a marked off position can be used as the disconnecting means if other disconnecting means are installed as follows:

1. In multifamily family dwellings, this disconnecting means must be installed at least on the same floor as the heater. It can control other heaters and lamps.
2. In two-family dwellings this disconnecting means can be installed either inside or outside the dwelling unit. There is also a provision for additional control.
3. In one-family dwellings the service disconnect can serve this purpose.
4. In other occupancies, if readily accessible, the branch-circuit switch or circuit breaker can be used.

424.20 Thermostatically Controlled Switching Devices

These devices can also serve as the disconnecting means if all the following conditions are met; have a marked off position, interrupt all ungrounded conductors when manually put in that position, cannot be automatically energized in the off position, and are located in accordance with Section 424.19.

424.21 Switch and Circuit Breaker to Be Indicating

424.22 Overcurrent Protection

Branch-circuit protection as provided in Article 210 can be used. Motors must comply with Articles 430 and 440. Resistance element cannot be protected at a higher rating than 60 A. Equipment rated at more than 48 A must be subdivided, with each subdivision rated not more than 48 A. The manufacturers must provide the division and overcurrent protection for the division. Conductors that supply supplementary overcurrent devices are branch circuit conductors. There are notations concerning heaters over 50 kW.

Field conductors between the heater and supplementary overcurrent protection must be sized at 125% of the load if the heater is rated less than 50 kW.

If the heater is rated at least 50 kW the conductors can be sized at 100% of the rating if the heater is marked for 100% conductor rating and the conductors are not smaller than the marked rating. Also, a temperature control device must be installed.

IV. Marking of Heating Equipment

424.28 Namplate

Must have proper marking information: volts, amperes, horsepower, etc.

424.29 Marking of Heating Elements

Volts and amperes or volts and watts.

V. Electric Space Heating Cables

424.34 Heating Cable Construction

Nonheating cables of at least 2.1 m (7 ft.) must be installed at the factory.

424.35 Marking of Heating Cables

Proper marking is required. Special color coding is required for different voltages of heating cables: 120 V, yellow; 208 V, blue; 240 V, red; 277 V, brown.

424.36 Clearances of Wiring in Ceilings

Fifty millimeters (two inches) of space is required above a heated ceiling. An operating ambient temperature of 50°C is to be used for all sizing and appropriate correction factors used. If located above thermal insulation 50 mm (2 in.) in thickness, the temperature correction is not needed.

424.37 Location of Branch-Circuit and Feeder Wiring in Walls

Comply with Article 300 and Section 310.10.

424.38 Area Restrictions

The cables cannot go beyond the room in which they originate. They cannot be installed in closets or over a partition that extends to be ceiling. There is also a restriction as to how they may be run over cabinets. They may be used in closet ceilings for low-temperature sources for humidity control, if there are no obstructions between floor and ceiling in that part of the ceiling.

424.39 Clearance from Other Objects and Openings

The heating cables must be at least 200 mm (8 in.) from boxes used for surface-mounted lighted luminaires (fixtures) and 50 mm (2 in.) from recessed lighting luminaires (fixtures). They cannot be installed where they are covered by any surface-mounted unit.

424.40 Splices

The cable's length cannot be changed.

424.41 Installation of Heating Cables on Dry Board, in Plaster and on Concrete Ceilings

(A) Cables are not permitted in walls unless it is a single run to reach a dropped ceiling.

(B) A minimum spacing of 38 mm (1½ in.) is required for cables rated at 9 watts/m (2¾ W per foot) and under for adjacent runs.

(C) Heating cables cannot be applied directly to metal surfaces. They can be only to fire-resistant surfaces. A coat of plaster must be applied to a metal surface.

(D) The entire heating cable, at least 75 mm (3 in.) of the nonheating cable and the splice between them must be embedded in plaster or dry board.

(E) The surface of the ceiling must have a coating of at least 13 mm (½ in.) of plaster or other approved material.

(F) Cables have to be secured at a maximum of 400 mm (16 in.) on center unless approved for 1.8 m (6 ft.) on center.

(G) When dry board is used for the ceiling, it cannot be thicker than 13 mm (½ in.).

(H) Cables cannot come in contact with metal or other surfaces which can conduct electricity.

(I) The cables must be installed parallel to joists in a dry board installation and at least 65 mm (2½ in.) between center of adjacent runs of cable under the center line of the joist. Care must be taken not to drive nails through the heating cables.

(J) The cables can cross joists at the end of the room or to comply with a manufacturer's requirement.

424.42 Finished Ceilings

Coverings cannot have thermal insulating properties. Paint and wallpaper can be used.

424.43 Installation of Nonheating Leads of Cables and Panels

424.44 Installation of Cables in Concrete or Poured Masonry Floors

424.45 Inspection and Tests

VI. Duct Heaters

424.57 General

424.58 Identification

424.59 Airflow

424.60 Elevated Inlet Temperature

Duct heaters must be approved for mounting in ducts and at the expected temperatures, as well as assure proper air-flow.

424.61 Installation of Duct Heaters with Heat Pumps and Air Conditioning

424.62 Condensation

424.63 Fan Circuit Interlock

424.64 Limit Controls

424.65 Location of Disconnecting Means

Must be within sight of the controller except as noted in Section 424.19(A).

424.66 Installation

VII. Resistance-Type Boilers

424.70 Scope

424.71 Identification

424.72 Overcurrent Protection

424.73 Over-Temperature Limit Control

424.74 Over-Pressure Limit Control

VIII. Electrode-Type Boilers

424.80 Scope

424.81 Identification

424.82 Branch-Circuit Requirements

424.83 Over-Temperature Limit Control

424.84 Over-Pressure Limit Control

424.85 Grounding

424.86 Markings

IX. Electric Radiant Heating Panels and Heating Panel Sets

424.90 Scope

424.91 Definition

424.92 Markings

424.93 Installation

424.94 Clearances of Wiring in Ceilings

424.95 Location of Branch-Circuit and Feeder Wiring in Walls

424.96 Connection to Branch-Circuit Conductors

424.97 Nonheating Leads

424.98 Installation in Concrete or Poured Masonry

424.99 Installation under Floor Covering

ARTICLE 426
FIXED OUTDOOR ELECTRIC DE-ICING AND SNOW-MELTING EQUIPMENT

I. General

426.1 Scope

This article covers installations both embedded and exposed.

426.2 Definitions

426.3 Application of Other Articles

The rest of the *NEC*® is applicable unless changed by this article.

426.4 Branch-Circuit Sizing

Conductors and overcurrent protection must be at least 125% of the load. Section 240.3(B) rating or setting is permitted.

II. Installation

426.10 General

426.11 Use

426.12 Thermal Protection

Must be protected against contact by persons if operated at temperatures above 60°C (140°F).

426.13 Identification

Warning signs must be posted where used.

426.14 Special Permission

III. Resistance Heating Elements

426.20 Embedded De-icing and Snow-Melting Equipment

Units cannot exceed 1300 watts/m^2 (120 W per square foot) and must have at least 25 mm (1 in.) between cables. They can be installed either on an asphalt or masonry base at least 50 mm (2 in.) thick and covered with at least 38 mm (1½ in.) of asphalt or masonry, or on any approved type of base embedded in 90 mm (3½ in.) of asphalt or masonry and covered with at least 38 mm (1½ in.) of asphalt or masonry. They must be secured and not bridge expansion joints unless proper provisions are made.

426.21 Exposed De-icing and Snow-Melting Equipment

They must be secured, proper temperature limiting, and have no cross-bridge expansion joints unless provided for.

426.22 Installation of Nonheating Leads for Embedded Equipment

426.23 Installation of Nonheating Leads for Exposed Equipment

426.24 Electrical Connection

426.25 Marking

426.26 Corrosion Protection

426.27 Grounding Braid or Sheath

426.28 Equipment Protection

IV. Impedance Heating

426.30 Personnel Protection

426.31 Isolation Transformer

426.32 Voltage Limitations

The secondary winding of the isolation transformer must not operate at greater than 30 V ac unless protected by a ground fault circuit interrupter. If ground fault circuit interrupter is used it can be greater than 30 V but not more than 80 V.

426.33 Induced Currents

426.34 Grounding

V. Skin Effect Heating

426.40 Conductor Ampacity

426.41 Pull Boxes

426.42 Single Conductor in Enclosure

426.43 Corrosion Protection

426.44 Grounding

VI. Control and Protection

426.50 Disconnecting Means

Must have a disconnecting means that opens the ungrounded wires. The branch-circuit breaker can be used if readily accessible. Also permitted is a factory-installed attachment plug if it is rated not more than 20 A and 150 V to ground.

426.51 Controllers

If the temperature controller operates line current and has an off position, it must open all ungrounded conductors. It cannot be used as a disconnecting means unless it can be locked in the open position. If no off position is provided, it does not have to open all ungrounded conductors and cannot be used as a disconnecting means. If it controls a relay, it does not have to open all ungrounded conductors and cannot be used as a disconnecting means. There are also rules in the *NEC*® for combined switching devices.

426.52 Overcurrent Protection

Refer to Section 426.4.

426.54 Cord- and Plug-Connected De-icing and Snow Melting Equipment

ARTICLE 427
FIXED ELECTRIC HEATING EQUIPMENT FOR PIPELINES AND VESSELS

I. General

427.1 Scope

427.2 Definitions

427.3 Application of Other Articles

427.4 Branch-Circuit Sizing

II. Installation

427.10 General

427.11 Use

427.12 Thermal Protection

427.13 Identification

III. Resistance Heating Elements

427.14 Secured
427.15 Not in Direct Contact
427.16 Expansion and Contraction
427.17 Flexural Capability
427.18 Power Supply Leads
427.19 Electrical Connections
427.20 Marking
427.22 Equipment Protection
427.23 Grounded Conductive Covering

IV. Impedance Heating

427.25 Personnel Protection
427.26 Isolation Transformer
427.27 Voltage Limitations
427.28 Induced Currents
427.29 Grounding
427.30 Secondary Conductor Sizing

V. Induction Heating

427.35 Scope
427.36 Personnel Protection
427.37 Induced Current

VI. Skin Effect Heating

427.45 Conductor Ampacity
427.46 Pull Boxes
427.47 Single Conductor Enclosure
427.48 Grounding

VII. Control and Protection

427.55 Disconnecting Means
427.56 Controls
427.57 Overcurrent Protection

ARTICLE 430
MOTORS, MOTOR CIRCUITS, AND CONTROLLERS

I. General

430.1 Scope

This article covers requirements for motors, motor branch circuit and feeder conductors and their protection, motor overload protection, motor control circuits, motor controllers, and motor control centers. There are general and specific requirements. Figure 430.1 in the *NEC*® describes where in Article 430 different parts of the circuits are covered as well as other articles. This article does not cover air-conditioning and refrigeration equipment, which is covered in Article 440 and motor control centers which is covered in 110.26(F).

430.2 Adjustable-Speed Drive Systems

430.3 Part Winding Motors

430.5 Other Articles

There are numerous other articles with which motor and controllers must comply for special cases. These are noted in Table 430.5.

430.6 Ampacity and Motor Rating Determination

Generally, the conductor size is determined by using the allowable ampacity tables as noted in Section 310.15(B) or calculated as noted in Section 310.15(C). Use Section 400.5 to determine flexible cord conductor size.

(A) General Motor Applications. Tables 430.147 through 430.150 are used to determine ampere rating for sizes of wire, switches, circuit protection, and the like (not the nameplate ratings). If the motor has a marking in amperes but not horsepower, it is assumed that the horsepower ratings correspond to those in Tables 430.147 through 150. There are three exceptions. Separate motor overload protection is based on the nameplate reading.

(B) Torque Motors. The rated current to be used for these motors is the locked-rotor current. This nameplate value is used to determine the size of the branch circuit conductors as noted in Sections 430.22 and 430.24, the rating of the overload protection, and the rating of the motor overload protection and the rating of the branch circuit short circuit and ground fault protection.

(C) Alternating Current Adjustable Voltage Motors. For these types of motors use the maximum operating current marked on the motor. If there is no marking, then use 150% of the values in Tables 430.149 and 430.150.

430.7 Marking on Motors and Multimotor Equipment

430.8 Marking on Controllers

430.9 Terminals

430.10 Wiring Space in Enclosures

Table 430.10(B) indicates the minimum wire bending space at the terminals of enclosed motors.

430.11 Protection against Liquids

430.12 Motor Terminal Housings

Tables 430.10(B), 430.12(B), 430.12(C)(1), and 430.12(C)(2) give various dimension and space requirements. There is also information on connections.

430.13 Bushing

430.14 Location of Motors

430.16 Exposure to Dust Accumulations

430.17 Highest-Rated or Smallest-Rated Motor

If this is to be determined for compliance with Sections 430.24, 430.53(B), and 430.53(C), use the rated full-load current from Tables 430.147, 430.148, 430.149, and 430.150.

430.18 Nominal Voltage of Rectifier Systems

II. Motor Circuit Conductors

430.21 General

The rules of this part apply to circuits supplying motors at 600 V or less.

430.22 Single Motor

Generally size conductors at 125% of the motor full-load rated current as noted in Section 430.6(A)(1) for the supply to a single motor. There are two exceptions to this rule. The first is for direct-current motors that operate from a single-phase power supply which is rectified. The second is for conductors supplying power conversion equipment. There are regulations for multispeed motors and wye-start delta-run motors and part winding motors. When the junction box is allowed to be separated from the motor, a special exception is made for wire size on motors of 1 hp or less. There are provisions for other than continuous-duty motors.

430.23 Wound-Rotor Secondary

The secondary leads for a continuous-duty wound-rotor motor must be rated at least 125% of the full-load secondary current. If the motor is not continuous, Table 430.22(E) must be used. Table 430.23(C) is used when the secondary resistor is separate from the controller.

430.24 Several Motors or a Motor(s) and Other Load(s)

In this case the conductors are rated at 125% of the full load current of the largest motor (the same as a single motor) plus the full-load current of all additional motors plus the ampere rating of the other loads which are calculated in accordance with Article 430.6(A) and other sections of the *NEC®*. There are three exceptions: continuous-duty, motor-operated fixed space heating, and interlocked motors.

430.25 Multimotor and Combination-Load Equipment

The conductors for these loads are based, as a minimum, on the circuit ampacity marked on the equipment in accordance with Section 430.7(D). If the equipment is not factory wired, and the nameplates visible as per Section 430.7(D)(2), the conductors are calculated in accordance with Section 430.24.

430.26 Feeder Demand Factor

If all motors will not operate simultaneously for any number of reasons, the authority having jurisdiction can allow the use of a demand factor.

430.27 Capacitors with Motors

430.28 Feeder Taps

Feeder tap conductors must be sized at least according to Part II of this section and either be enclosed and 3.0 m (10 ft.) or less in length or be one-third or more of the rating of the feeder, 7.5 m (25 ft.) or less in length, and protected. There is an exception for feeder taps over 7.5 m (25 ft.) in high bay manufacturing buildings. There are additional rules.

430.29 Constant-Voltage DC Motors—Power Resistors

III. Motor and Branch-Circuit Overload Protection

430.31 General

The overload protection is usually the thermal overloads in the controller or starter. It should not be confused with the short-circuit protection, which is the fuse or circuit breaker in the panelboard. Diagram 430.1 in the Code should be referred to, as well as Annex D, Example No. D8.

430.32 Continuous-Duty Motors

There are four methods acceptable:

(A) More than 1 Horsepower

1. A separate overload device must be used. It must be rated at not more than 125% of the motor nameplate rating if the motor is rated with a temperature rise of not more than 40°C or a service factor of 1.15 or more. The rating is reduced to 115% for all other motors. Exceptions are provided in Section 430.32(C).
2. Thermal protectors as part of the motor are listed in this paragraph.
3. If the motor is part of an approved assembly, will not normally have overloads, and has some protection against failure to start, additional overload protection is not necessary.
4. Motors larger than 1500 hp are discussed in this paragraph.

(B) One Horsepower or Less, Automatically Started. Here again there are four acceptable methods; the first three are similar to the first three for motors rated more than 1 hp. These include separate devices (ratings are indicated), thermal protectors, and cases where no additional protection other than branch-circuit protection is required (where protection is integral with the motor). A fourth case is that where the impedance of the motor protects against overheating when the motor does not start.

(C) Selection of Overload Relay. In the event the ratings chosen as listed in Section 430.32(a)(1) and (c)(1) will not allow the motor to start, this section gives larger permissible ratings.

(D) One Horsepower or Less, Nonautomatically Started. If it is within sight of the controller location, the ground-fault protection and branch-circuit chort-circuit protection is sufficient. The sizings noted in Part IV are to be used. It can be on a 120-V circuit rated 20 A or less. If not within sight of the controller or permanently installed, it must comply with Section 430.32(B).

(E) Wound Rotor Secondaries. These can be protected by the motor-overload device.

430.33 Intermittent and Similar Duty

You can omit the separate overload protection for these types of motors and use only the branch-circuit, short-circuit and ground-fault protective device.

430.35 Shunting during Starting Period

There are cases where the overload protection can be shunted during starting.

430.36 Fuses—In Which Conductor

When used as overload protection, put one in each ungrounded conductor and also the grounded conductor if the supply is three-phase, three-wire ac with one phase grounded.

430.37 Devices Other than Fuses—In Which Conductor

Table 430.37 gives the requirements.

430.38 Number of Conductors Opened by Overload Devices

The number of conductors to be opened must prevent current to the motor for devices other than fuses or thermal protection.

430.39 Motor Controller as Overload Protection

This is permitted if it is in accordance with Table 430.37 and operator in the running position and also in the starting position for an ac motor.

430.40 Overload Relays

If the device used for motor overload protection cannot clear short circuits, they must be protected by either fuses or circuit breakers or a motor short-circuit protector sized as noted in Section 430.52. There is one exception to the sizing rule—group installations and where an overload relay is used.

430.42 Motors on General-Purpose Branch Circuits

This section indicates what overload protection must be used when motors are connected to general-purpose branch circuits. It must also comply with Article 210.

(A) Not over 1 Horsepower. Can be connected without overload protection if the rules of Section 430.32(B) and (D) and Section 430.53(A)(1) and (A)(2) are followed.

(B) Over 1 Horsepower. Each motor must have overload protection as noted in Section 430.32.

(C) Cord- and Plug-Connected. If no overload protection is provided, as in (a) above, the ratings of the plug and receptacle cannot be larger than 15 A 125 V or 10 A 250 V. If protection is required as in (b) above, the overload device must be part of the motor or appliance.

(D) Time Delay. Time delay must be provided in the branch-circuit and ground-fault devices to let the motor start.

430.43 Automatic Restarting

430.44 Orderly Shutdown

If an orderly shutdown is necessary to protect people rather than a quick opening of the circuit, the overload device can be connected to an alarm rather than open the circuit under the provisions of this section.

IV. Motor Branch-Circuit, Short-Circuit, and Ground-Fault Protection

430.51 General

430.52 Rating or Setting for Individual Motor Circuit

The device must carry the starting current. Table 430.152 lists the maximum settings of the protective devices. If the values in the tables are not a standard-size device, the rating can be increased to the next highest rating. If the rating in the table will not permit the motor to start, the rating can be increased as follows:

1. Nontime-delay fuse less than 601 A up to 400% of full-load current
2. Time-delay fuses up to 225% of full-load current
3. Circuit breakers (inverse time) up to 400% for full-load currents of up to 100 A and 300% for larger currents
4. Fuses between 601 and 6000 A up to 300% of full-load current

Also listed in this section are the rules for when instantaneous trip circuit breakers and motor short-circuit protectors can be used; rules for multispeed motors; multispeed motor; power electronic devices; rules for torque motors; and rules for phase converters. Where a maximum rating is indicated by the manufacturer, it cannot be exceeded. The reader is referred to this section of the Code for complete information.

430.53 Several Motors or Loads on One Branch Circuit

This section indicates under what conditions two or more motors or combined loads are permitted on one branch circuit.

(A) If none of the motors is greater than 1 hp, the circuit is protected at 20 A at 120 V or not more than 15 A at 600 V, the full-load current of each motor is not more than 6 A, individual overload protection meets the requirements of Section 430.32, and the rating marked on any controller for branch-circuit short circuit and ground-fault protection is not exceeded, more than one motor or combination loads are permitted on one branch circuit.

(B) When the branch-circuit short-circuit and ground-fault protective devices rating is not larger than required for the smallest motor and all motors have individual overload protection and the combination load will operate under these conditions, more than one motor or combination loads are permitted.

(C) Group installations of more than one motor or combination loads are permitted if the overload devices and motor controllers are in a listed factory assembly, and motor short circuit and ground-fault protection are in the assembly or noted on the assembly; or if the controllers, overload devices, branch circuit, short circuit, and ground-fault devices are field-installed separately with manufacturer's instructions for use with each other and the following conditions are met:

1. The overload device must be approved for group installation and have a specified maximum fuse or inverse time circuit breaker rating, or both.
2. The motor controller for each motor must comply with the same rules as in (1) above.
3. The circuit breakers must be listed for group installation and have inverse time characteristics.
4. The rating of the fuse or circuit breaker for the branch circuit must comply with rules of Sections 430.52 and 240.3(B). It must be at least that of Section 430.52 for the largest motor and the sum of full-load ratings of other motors and loads.
5. The rating allowed in Section 430.40 for the smallest motor will also be the maximum for the branch-circuit fuses or inverse time circuit breakers.

(D) In a group installation as described in (c) above, if the conductors to any single motor are not larger than the branch-circuit conductors or are not less than one-third the rating of the branch-circuit conductors in accordance with Section 430.22, and not farther than 7.5 m (25 ft.) from the controller and physically protected, then installation will not require individual branch-circuit short-circuit and ground-fault protection.

430.54 Multimotor and Combination-Load Equipment

The branch-circuit short-circuit and ground-fault device rating must not be larger than the rating marked on the equipment and comply with Section 430.7(D).

430.55 Combined Overcurrent Protection

As long as the rating is not larger than that required for overload protection in Section 430.32, the branch-circuit short-circuit and ground-fault protection and overload protection can be in one device.

430.56 Branch-Circuit Protective Devices—In Which Conductor

430.57 Size of Fuseholder

430.58 Rating of Circuit Breaker

V. Motor Feeder Short-Circuit and Ground-Fault Protection

430.61 General

430.62 Rating or Setting—Motor Load

The rating must be that of the largest branch-circuit protective device plus the full-load currents of all other motors. If larger feeders have been installed for future expansion, the rating can be that of the feeder. Reference is made to Sections 430.52 and 440.22(A). There are two exceptions and additional requirements.

430.63 Rating or Setting—Power and Light Loads

When combined loads are installed in one feeder, the rating can be sufficient to carry the lighting and appliance lead plus that permitted by Section 430.52 for a single motor and Section 430.62 for more than one motor and 440.22 for a single hermetic refrigerant motor-compressor. There is an excpetion for a motor control center.

VI. Motor Control Circuits

430.71 General

The motor control circuit carries the electric signal to control the controller but not power for the circuit.

430.72 Overcurrent Protection

If the circuit is tapped from the load side of a branch-circuit protective device, it must have protection as specified in Section 430.72. If not, it must be protected as specified in Section 725.23. There are two exceptions to this rule.

There are also requirements for a control circuit transformer and additional requirements for conductor protection.

430.73 Mechanical Protection of Conductor

430.74 Disconnection

Motor control circuits must have a disconnecting means. One device can disconnect both the motor and control circuits from power, or separate ones can be used. If a transformer is used to reduce the voltage for the control circuit and it is located in the controller enclosure, the disconnecting means must be in the supply side of the transformer. There are a number of exceptions.

VII. Motor Controllers

430.81 General

The word "controller" is defined in Article 100. Here it can mean a device or switch to stop and start a motor by breaking the motor circuit current. When the motor is stationary and of not more than ⅛ hp, the branch-circuit protective device can be used if the motor is normally left running and cannot be damaged by not starting or by overloads. The plug and receptacle can be the controller if the motor is portable and not more than ⅓ hp.

430.82 Controller Design

430.83 Ratings

Generally, it must have a horsepower rating at least equal to that of the motor. There are a number of additional rules.

430.84 Need Not Open All Conductors

Unless it serves as a disconnecting device, it does not have to open all conductors.

430.85 In Grounded Conductors

This is permitted if it also simultaneously opens all other conductors.

430.87 Number of Motors Served by Each Controller

Each motor must normally have its own controller. There are exceptions to this rule if the motors are rated at not more than 600 V and the single controller is rated at not less than the sum of all motors it will control. Under these conditions it is allowed if more than one motor is used in one device or one overcurrent device is used for a group of motors [Section 430.53(A)] or the motors are in one room in sight of the controller.

430.88 Adjustable-Speed Motors

430.89 Speed Limitation

430.90 Combination Fuseholder and Switch as Controller

430.91 Motor Controller Enclosure Types

VIII. Motor Control Centers

430.92 General

A motor control center is an assembly of enclosed sections which principally contain motor control units and have a common power bus.

430.94 Overcurrent Protection

They must have overcurrent protection, either an overcurrent device upstream or a main device in the motor control center. It must be in accordance with Parts A, B and I of Article 240 and sized in accordance with the common power bus.

430.95 Service-Entrance Equipment

430.96 Grounding

All sections must be bonded with an equipment grounding conductor or grounding bus in accordance with Table 250.95. All equipment grounding conductors must terminate here. In the case of a single section a grounding termination point must be provided.

430.97 Busbars and Conductors

(A) Support and Arrangement. They must be protected from damage and held firmly. Only the conductors connected or used for control are allowed in a section.

(B) Phase Arrangement. Three phase buses are arranged A, B, C, left to right, top to bottom, or front to back when standing in the front.

(C) Minimum Wire Bending Space. As noted in Article 373.

(D) Spacings. As noted in Table 430.97.

(E) Barriers. Used in service-entrance motor control centers.

430.98 Marking

IX. Disconnecting Means

430.101 General

430.102 Location

It must be provided for and in sight from the controller location. An exception is made for motors over 600 V if it can be locked out and a warning sign is posted. Also, one disconnect can be used for a group of controllers for a multimotor device. A disconnecting means must also be located within sight of the motor location except in accordance with Section 430.102(a). There are some additional exceptions.

430.103 Operation

430.104 To Be Indicating

430.105 Grounded Conductors

If all conductors are opened simultaneously, one pole can be in a grounded conductor.

430.107 Readily Accessible

430.108 Every Switch

430.109 Type

The disconnecting device must be noted in (A) below unless permitted in (B) through (G) meeting the specified conditions.

(A) General. It must be a horsepower-rated motor circuit switch. A motor circuit switch for a Design E motor 2 hp or more must be marked for a Design E motor and it must have a horsepower rating 1.4 times or more than the rating of a 3- through 100-hp motor or 1.3 times or more the rating of a motor larger than 100 hp. It can also be a molded case circuit breaker, molded case switch, an instantaneous trip circuit breaker which is part of a combination motor controller, a self-protected combination controller, or a manual motor controller if it is marked as such, if it is installed between the motor and the final motor branch circuit and ground-fault protective device, all of the items being listed.

(B) Stationary Motors of 1/8 Horsepower or Less. The branch circuit overcurrent device can be used.

(C) Stationary Motors of 2 Horsepower or Less. Stationary motors 2 hp or less and 300 V or less can have a general-use switch rated twice that of the motor. On ac circuits a general-use snap switch listed for ac use only can be used if the motor is not more than 80% of the rating of the switch. A listed manual motor controller can also be used if it is marked as being suitable as a motor disconnect and it is horsepower rated not less than the rating of the motor.

(D) Autotransformer-Type Controlled Motors. Motors over 2 hp up to 100 hp can use a general-use switch for motors with autotransformer controllers if the motor drives a generator with overload protection; the controller can interrupt locked rotor current, has no voltage release, and has running overload protection not larger than 125% of full-load current; separate fused or inverse time circuit breakers are provided and rated not more than 150% of full load current.

(E) Isolating Switches. Stationary motors more than 40 hp dc or 100 hp ac can have a general use or isolating switch if marked "Do Not Operate Under Load."

(F) Cord- and Plug-Connected Motors. A horsepower rated attachment plug and receptacle rated at least the rating of the motor can be use for other than Design E motors rated more that 2 hp. There are requirements for a Design E motor rated more than 2 hp. This is not required for cord and plug appliances (Section 422.32), room air conditioners (Section 440.63) or a portable motor 1/3 hp or less.

(G) Torque Motors. A general-use switch can be used for a torque motor.

430.110 Ampere Rating and Interrupting Capacity

(A) It must have a rating of at least 115% of the full-load current rating of the motor.

(B) It must have a rating of at least 115% of the nameplate current for torque motors.

(C) For combination loads of motors or motors and other types of loads, the ratings of the disconnect must not be less than 115% of the total of the full-load current of all the loads. The Code makes reference to the method of determining the combined load.

430.111 Switch or Circuit Breaker as Both Controller and Disconnecting Means

A switch or circuit breaker can be used as the disconnect and controller if it complies with Section 430.83: if it opens all ungrounded conductors to the motor, has an overcurrent protective device which opens all ungrounded conductors to the switch, and is either an air-break switch, an inverse time circuit breaker, or an oil switch.

430.112 Motors Served by Single Disconnecting Means

Generally, a motor must have its own disconnecting means. There is an exception.

430.113 Energy from More than One Source

Generally, each source must have a disconnect. There are two exceptions.

X. Over 600 Volts, Nominal

430.121 General

430.122 Marking on Controllers

430.123 Conductor Enclosures Adjacent to Motors

430.124 Size of Conductors

430.125 Motor-Circuit Overcurrent Protection

430.126 Rating of Motor Control Apparatus

430.127 Disconnecting Means

XI. Protection of Live Parts—All Voltages

430.131 General

430.132 Where Required

If operating at 50 V or more, the exposed live parts must be guarded against accidental contact by enclosure or location. This can be done in a special room or enclosure, on a balcony or by installation at least 2.5 m (8 ft.) above the floor. There is an exception.

430.133 Guards for Attendants

If the only guards are by location, as noted in Section 430.132, then insulating mats or platforms are required.

XII. Grounding—All Voltages

430.141 General

430.142 Stationary Motors

The frames must be grounded if in a wet location and not isolated, supplied by wire in a metal enclosure, in a hazardous location, or if motor operates over 150 V to ground.

430.143 Portable Motors

430.144 Controllers

430.145 Method of Grounding

XIII. Tables

ARTICLE 440
AIR CONDITIONING AND REFRIGERATING EQUIPMENT

I. General

440.1 Scope

This section states the scope.

440.2 Definitions

440.3 Other Articles

Many additional articles are listed, such as 422, 424, 430, 460, and 470, and some not in Chapter 4.

440.4 Marking on Hermetic Refrigerant Motor-Compressors and Equipment

440.5 Marking on Controllers

440.6 Ampacity and Rating

The rating on the nameplate is to be used for a hermetic refrigerant motor compressor. There are two exceptions. Conductors are to be chosen from Tables 310.16 through 310.19 or in accordance with Section 310.15. Requirements are noted for multimotor equipment.

440.7 Highest Rated (Largest) Motor

440.8 Single Machine

A refrigerating system or air-conditioning system can be considered in accordance with Section 430.87, Exception, and Section 430.112, Exception.

II. Disconnecting Means

440.11 General

440.12 Rating and Interrupting Capacity

For hermetic refrigerant motor-compressor it must be at least 115% of the larger of the rated load or branch-circuit election circuit. There is an exception. If this rating comes out to be 100 hp or larger, use the provisions of Section 430.109(E). Information is given concerning equivalent horsepower and locked rotor current. Additional information is given concerning combination loads, horsepower rating, full load equivalent, small motor compressors and switches.

440.13 Cord-Connected Equipment

The separable connector or plug and receptacle can be used as the disconnecting means.

440.14 Location

It must be within sight and readily accessible to the equipment. There are two exceptions.

III. Branch-Circuit Short-Circuit and Ground-Fault Protection

440.21 General

440.22 Application and Selection

(A) Rating or Setting for Individual Motor-Compressor. It must allow the motor to start. The *NEC*® therefore permits a rating or setting of up to 175% of the larger of the motor load or branch-circuit selection current. This can be increased to 225% if necessary. A 15-A rating is the least that can be required for branch-circuit, short-circuit and ground-fault protection.

(B) Rating or Setting for Equipment. It must also allow the equipment to start. If only one hermetic refrigerant motor compressor is used, conform to Section 422(A). If more than one hermetic refrigerant motor compressor is used, conform to Section 430.53 and the following:

(1) If the hermetic refrigerant motor compressor is the largest load, use Section 440.22(A) for this plus the other loads.

(2) If the hermetic refrigerant motor compressor is not the largest load, the sum is that of the rating of the motor compressor loads plus the value noted in Section 430.53(C)(4) or Section 240.3 for other loads, whichever is applicable. There are two exceptions. For a detailed analysis, refer to the *NEC*® and one of the many reference guides.

(C) If a maximum rating is indicated by the manufacturer, it should not be exceeded.

IV. Branch-Circuit Conductors

440.31 General

440.32 Single Motor-Compressor

The rating cannot be less than 125% of the largest of the motor compressor rated load or branch-circuit selection current. There is additional information concerning wye-start, delta-run motor compressor.

440.33 Motor-Compressor(s) with or without Additional Motor Loads

The ampacity must be the sum of the larger of the rated load or branch-circuit selection currents of all the motor-compressor loads plus full-load currents of the other motors plus 25% of the largest motor. There are two exceptions to this: one is for room air conditioners as per Part VII and the other is for interlocked motors.

440.34 Combination Load

The capacity must be adequate for all other loads, plus the motor-compressor load as specified in Sections 430.32 and 440.33. An exception is also made for interlocked circuitry.

440.35 Multimotor and Combination-Load Equipment

It cannot be less than that marked on the equipment as per Section 440.4(B).

V. Controller for Motor-Compressors

440.41 Rating

Generally, use the nameplate ratings. If more than one load is served, use the combined load rating [refer to Section 440.12(B)]. There are additional requirements.

VI. Motor Compressor and Branch-Circuit Overload Protection

440.51 General

440.52 Application and Selection

(A) Protection of Motor-Compressor. The following methods can be used: (1) an overload relay that will trip at not more than 140% of motor full-load current; (2) a thermal protector integral with the motor compressor; (3) a fuse or inverse time circuit breaker responsive to motor current and rated not more than 125% of motor current; and (4) a protective system furnished with the unit or specified for the unit.

(B) Protection of Motor-Compressor Control Apparatus and Branch-Circuit Conductors. It can be the same device protecting the motor-compressor in accordance with Section 440.52(A). There is an exception.

440.53 Overload Relays

440.54 Motor-Compressors and Equipment on 15- or 20-Ampere Branch Circuits—Not Cord and Attachment Plug-Connected

440.55 Cord and Attachment Plug-Connected Motor-Compressors and Equipment on 15- or 20-Ampere Branch Circuits

VII. Provisions for Room Air Conditioners

440.60 General

440.61 Grounding

440.62 Branch-Circuit Requirements

The branch circuit can have a load of only 80% of its rating for a cord- and plug-connected air-conditioning unit, if no other loads are on the circuit. If other loads are on the circuit, then the load of the cord- and plug-connected air-conditioning unit cannot be more than 50% of the rating of the branch circuit.

440.63 Disconnecting Means

The attachment plug can be used for a single-phase unit at 250 Volts or less if the manual control on the unit is readily accessible and within 1.8 m (6 ft.) of the floor or if a manual switch is readily accessible within sight of the unit.

and equipment. There are also additional requirements for storage batteries, generator set, uninterruptible power supplies, separate service, and unit equipment.

IV. Emergency System Circuits for Lighting and Power

700.15 Loads on Emergency Branch Circuits

Only those appliances and lamps needed for emergency use can be connected to the emergency lighting circuits.

700.16 Emergency Illumination

700.17 Circuits for Emergency Lighting

700.18 Circuits for Emergency Power

V. Control—Emergency Lighting Circuits

700.20 Switch Requirements

700.21 Switch Location

700.22 Exterior Lights

VI. Overcurrent Protection

700.25 Accessibility

700.26 Ground-Fault Protection of Equipment

This is not required for the alternate source. Ground fault indication is required in accordance with Section 700.7(D).

ARTICLE 701
LEGALLY REQUIRED STANDBY SYSTEMS

A. General

701.1 Scope

This article lays out the electrical safety requirements for the installation, operation, and maintenance of legally required standby power systems. This includes all equipment necessary to deliver power to the load. The reader is referred to NFPA 99-1999 (ANSI) Standard for Health Care Facilities for additional information, NFPA 110-1999 (ANSI) Emergency and Standby

Power Systems for performance information, and to ANSI/IEEE 446-1995 Recommended Practice for Emergency and Standby Power Systems for Industrial and Commercial Applications for further information.

701.2 Legally Required Standby Systems

These are systems that have this classification set by any governmental agency having jurisdiction. They are intended to supply power automatically to selected loads other than emergency system loads if the normal source fails. They differ from emergency systems in that they are not required for direct safety to life. They do supply power where the interruption might create a hazard or hamper rescue or firefighting operations. Some examples are given in the code to help the reader.

701.3 Application of Other Articles

Applicable articles of the Code apply unless otherwise modified in this Article.

701.4 Equipment Approval

Items must be approved for this use.

701.5 Test and Maintenance for Legally Required Standby Systems

(A) Conduct or Witness Test. When the installation is complete the authority having jurisdiction must either conduct or witness a test.

(B) Tested Periodically. To assure proper maintenance and operation the system must be tested on a schedule accepted by the authority having jurisdiction.

(C) Battery Systems Maintenance. Periodic maintenance is required if the batteries are used for control, starting, or ignition of a prime mover.

(D) Written Record. A written record must be kept of all tests and maintenance.

(E) Testing under Load. Provisions must be made for testing under load.

701.6 Capacity and Rating

The system must be able to supply the entire load operating together. It must also be able to withstand the available fault current.

701.7 Transfer Equipment

The transfer equipment must be automatic, identified for standby use, and approved by the authority having jurisdiction. Interconnection of the normal and emergency sources must be prevented. Bypass means for isolation of the transfer switch is permitted. Automatic transfer switches have to be electrically operated and mechanically held.

701.8 Signals

Audible and visual signals must be provided to indicate the derangement of the standby source, that the standby source is carrying the load, and that the battery charger is not functioning.

701.9 Signs

The type and location of the system must be indicated on a sign at the service entrance. There is an exception for certain unit equipment. If the system is grounded to a remote grounding electrode a sign must be placed at the grounding location identifying all emergency and normal sources connected there.

II. Circuit Wiring

701.10 Wiring Legally Required Standby Systems

The wiring for this system can occupy the raceways, cables, boxes, cabinets, etc. as the general wiring.

III. Sources of Power

701.11 Legally Required Standby Power Systems

The current must be available for use within 60 seconds of the loss of the normal power. The supply system can be any one of the following: storage battery, generator set, uninterruptible power supply, separate service, connection ahead of the service disconnecting means, and unit equipment. Each is covered in the code in detail as to specific requirements. Again it must be noted that some of these may not be permitted for specific occupancies as noted in other documents (i.e., health care facilities).

VI. Overcurrent Protection

701.15 Accessibility

701.17 Ground-Fault Protection of Equipment

This is not required for the alternate source.

ARTICLE 702
OPTIONAL STANDBY SYSTEMS

I. General

702.1 Scope

This article applies only to the installation and operation of optional standby systems.

702.2 Optional Standby Systems

These systems protect business or property where life safety is not dependent on the system. Power can be supplied either automatically or manually.

702.3 Application of Other Articles

Unless modified by this article, all provisions of the Code apply.

702.4 Equipment Approval

All equipment must be approved for this use.

702.5 Capacity and Rating

The system must be able to supply the entire load operating together. It must also be able to withstand the available fault current. The user can select the load which is connected to the system.

702.6 Transfer Equipment

Transfer equipment must be suitable for its use and designed and installed to prevent the inadvertent interconnection of the normal and alternate sources of power.

If it is located on the load side of branch-circuit protection, it can have supplementary overcurrent protection with an interrupting rating that can handle the generator available fault current.

702.7 Signals

Audible and visual signals must be provided to indicate the derangement of the standby source and that the standby source is carrying the load.

702.8 Signs

The type and location of the system must be indicated on a sign at the service entrance. There is an exception for certain unit equipment. If the system is grounded to a remote grounding electrode a sign must be placed at the grounding location identifying all optional standby power and normal sources connected there.

II. Circuit Wiring

702.9 Wiring Optional Standby Systems

The wiring for this system can occupy the raceways, cables, boxes, etc., as the general wiring.

702.10 Portable Generator Grounding

(A) Separately Derived System. Must be grounded to a grounding electrode in accordance with 250.30.

(B) Nonseparately Derived System. It must be bonded to the system grounding electrode.

The following article titles are listed for reference: **Article 705**—Interconnected Electric Power Production Sources; **Article 720**—Circuits and Equipment Operating at Less than 50 Volts; **Article 725**—Class 1, Class 2, and Class 3 Remote-Control, Signaling, and Power-Limited Circuits; **Article 727**—Instrumentation Tray Cable, Type ITC; **Article 760**—Fire Alarm Systems; **Article 770**—Optical Fiber Cables and Raceways; **Article 780**—Closed-Loop and Programmed Power Distribution.

CHAPTER 8
COMMUNICATIONS SYSTEMS

Article 800—Communications Circuits: **Article 810**—Radio and Television Equipment; **Article 820**—Community Antenna Television and Radio Distribution Systems; **Article 830**—Network-Powered Broadband Communications Systems.

9

TABLES

TABLES

TABLE 1 Percent of Cross Section of Conduit and Tubing for Conductors

Number of Conductors	*All Conductor Types*
1	53
2	31
Over 2	40

FPN No. 1: Table 1 is based on common conditions of proper cabling and alignment of conductors where the length of the pull and the number of bends are within reasonable limits. It should be recognized that, for certain conditions, a larger size conduit or a lesser conduit fill should be considered.

FPN No. 2: When pulling three conductors or cables into a raceway, if the ratio of the raceway (inside diameter) to the conductor or cable (outside diameter) is between 2.8 and 3.2, jamming can occur. While jamming can occur when pulling four or more conductors or cables into a raceway, the probability is very low.

NOTES TO TABLES

(1) See Appendix C for the maximum number of conductors and fixture wires, all of the same size (total cross-sectional area including insulation) permitted in trade sizes of the applicable conduit or tubing.

(2) Table 1 applies only to complete conduit or tubing systems and is not intended to apply to sections of conduit or tubing used to protect exposed wiring from physical damage.

(3) Equipment grounding or bonding conductors, where installed, shall be included when calculating conduit or tubing fill. The actual dimensions of the equipment grounding or bonding conductor (insulated or bare) shall be used in the calculation.

(4) Where conduit or tubing nipples having a maximum length not to exceed 600 mm (24 inches) are installed between boxes, cabinets, and similar enclosures, the nipples shall be permitted to be filled to 60 percent of their total cross-sectional area, and Section 310.15(B)(2)(a) need not apply to this condition.

(5) For conductors not included in Chapter 9, such as multiconductor cables, the actual dimensions shall be used.

(6) For combinations of conductors of different sizes, use Tables 5 and 5A for dimensions of conductors and Table 4 for the applicable conduit or tubing dimensions.

(7) When calculating the maximum number of conductors permitted in a conduit or tubing, all of the same size (total cross-sectional area including insulation), the next higher whole number shall be used to determine the maximum number of conductors permitted when the calculation results in a decimal of 0.8 or larger.

(8) Where bare conductors are permitted by other sections of this Code, the dimensions for bare conductors in Table 8 of Chapter 9 shall be permitted.

(9) A multiconductor cable of two or more conductors shall be treated as a single conductor for calculating percentage conduit fill area. For cables that have elliptical cross sections, the cross-sectional area calculation shall be based on using the major diameter of the ellipse as a circle diameter.

TABLE 4 Dimensions and Percent Area of Conduit and Tubing

(Areas of Conduit or Tubing for the Combinations of Wires Permitted in Table 1, Chapter 9)

Article 358—Electrical Metallic Tubing (EMT)

Metric Designator	Trade Size	Internal Diameter		Total Area 100%		2 Wires 31%		Over 2 Wires 40%		1 Wire 53%		60%	
		mm	*in.*	*mm²*	*in.²*	*mm²*	*in.²*	*mm²*	*in.²*	*mm²*	*in.²*	*mm²*	*in.²*
16	½	15.8	0.622	196	0.304	61	0.094	78	0.122	104	0.161	118	0.182
21	¾	20.9	0.824	343	0.533	106	0.165	137	0.213	182	0.283	206	0.320
27	1	26.6	1.049	556	0.864	172	0.268	222	0.346	295	0.458	333	0.519
35	1-¼	35.1	1.380	968	1.496	300	0.464	387	0.598	513	0.793	581	0.897
41	1-½	40.9	1.610	1314	2.036	407	0.631	526	0.814	696	1.079	788	1.221
53	2	52.5	2.067	2165	3.356	671	1.040	866	1.342	1147	1.778	1299	2.013
63	2-½	69.4	2.731	3783	5.858	1173	1.816	1513	2.343	2005	3.105	2270	3.515
78	3	85.2	3.356	5701	8.846	1767	2.742	2280	3.538	3022	4.688	3421	5.307
91	3-½	97.4	3.834	7451	11.545	2310	3.579	2980	4.618	3949	6.119	4471	6.927
103	4	110.1	4.334	9521	14.753	2951	4.573	3808	5.901	5046	7.819	5712	8.852

TABLE 4 (continued)

Article 362—Electrical Nonmetallic Tubing (ENT)

Metric Designator	Trade Size	Internal Diameter		Total Area 100%		2 Wires 31%		Over 2 Wires 40%		1 Wire 53%		60%	
		mm	in.	mm^2	$in.^2$	mm^2	$in.^2$	mm^2	$in.^2$	mm^2	$in.^2$	mm^2	$in.^2$
16	½	14.2	0.560	158	0.246	49	0.076	63	0.099	84	0.131	95	0.148
21	¾	19.3	0.760	293	0.454	91	0.141	117	0.181	155	0.240	176	0.272
27	1	25.4	1.000	507	0.785	157	0.243	203	0.314	269	0.416	304	0.471
35	1-¼	34.0	1.340	908	1.410	281	0.437	363	0.564	481	0.747	545	0.846
41	1-½	39.9	1.570	1250	1.936	388	0.600	500	0.774	663	1.026	750	1.162
53	2	51.3	2.020	2067	3.205	641	0.993	827	1.282	1095	1.699	1240	1.923
63	2-½	—	—	—	—	—	—	—	—	—	—	—	—
78	3	—	—	—	—	—	—	—	—	—	—	—	—
91	3-½	—	—	—	—	—	—	—	—	—	—	—	—

TABLE 4 (continued)

Article 348—Flexible Metal Conduit (FMC)

Metric Designator	Trade Size	Internal Diameter		Total Area 100%		2 Wires 31%		Over 2 Wires 40%		1 Wire 53%		60%	
		mm	in.	mm^2	$in.^2$	mm^2	$in.^2$	mm^2	$in.^2$	mm^2	$in.^2$	mm^2	$in.^2$
12	⅜	9.7	0.384	74	0.116	23	0.036	30	0.046	39	0.061	44	0.069
16	½	16.1	0.635	204	0.317	63	0.098	81	0.127	108	0.168	122	0.190
21	¾	20.9	0.824	343	0.533	106	0.165	137	0.213	182	0.283	206	0.320
27	1	25.9	1.020	527	0.817	163	0.253	211	0.327	279	0.433	316	0.490
35	1-¼	32.4	1.275	824	1.277	256	0.396	330	0.511	437	0.677	495	0.766
41	1-½	39.1	1.538	1201	1.858	372	0.576	480	0.743	636	0.985	720	1.115
53	2	51.8	2.040	2107	3.269	653	1.013	843	1.307	1117	1.732	1264	1.961
63	2-½	63.5	2.500	3167	4.909	982	1.522	1267	1.963	1678	2.602	1900	2.945
78	3	76.2	3.000	4560	7.069	1414	2.191	1824	2.827	2417	3.746	2736	4.241
91	3-½	88.9	3.500	6207	9.621	1924	2.983	2483	3.848	3290	5.099	3724	5.773
103	4	101.6	4.000	8107	12.566	2513	3.896	3243	5.027	4297	6.660	4864	7.540

TABLE 4 (continued)

Article 342—Intermediate Metal Conduit (IMC)													
Metric Designator	*Trade Size*	*Internal Diameter*		*Total Area 100%*		*2 Wires 31%*		*Over 2 Wires 40%*		*1 Wire 53%*		*60%*	
		mm	*in.*	*mm²*	*in.²*	*mm²*	*in.²*	*mm²*	*in.²*	*mm²*	*in.²*	*mm²*	*in.²*
12	⅜	—	—	—	—	—	—	—	—	—	—	—	—
16	½	16.8	0.660	222	0.342	69	0.106	89	0.137	117	0.181	133	0.205
21	¾	21.9	0.864	377	0.586	117	0.182	151	0.235	200	0.311	226	0.352
27	1	28.1	1.105	620	0.959	192	0.297	248	0.384	329	0.508	372	0.575
35	1-¼	36.8	1.448	1064	1.647	330	0.510	425	0.659	564	0.873	638	0.988
41	1-½	42.7	1.683	1432	2.225	444	0.690	573	0.890	759	1.179	859	1.335
53	2	54.6	2.150	2341	3.630	726	1.125	937	1.452	1241	1.924	1405	2.178
63	2-½	64.9	2.557	3308	5.135	1026	1.592	1323	2.054	1753	2.722	1985	3.081
78	3	80.7	3.176	5115	7.922	1586	2.456	2046	3.169	2711	4.199	3069	4.753
91	3-½	93.2	3.671	6822	10.584	2115	3.281	2729	4.234	3616	5.610	4093	6.351
103	4	105.4	4.166	8725	13.631	2705	4.226	3490	5.452	4624	7.224	5235	8.179

TABLE 4 (continued)

Article 356—Liquidtight Flexible Nonmetallic Conduit (LFNC-B*)

Metric Designator	Trade Size	Internal Diameter		Total Area 100%		2 Wires 31%		Over 2 Wires 40%		1 Wire 53%		60%	
		mm	*in.*	*mm²*	*in.²*	*mm²*	*in.²*	*mm²*	*in.²*	*mm²*	*in.²*	*mm²*	*in.²*
12	⅜	12.5	0.494	123	0.192	38	0.059	49	0.077	65	0.102	74	0.115
16	½	16.1	0.632	204	0.314	63	0.097	81	0.125	108	0.166	122	0.188
21	¾	21.1	0.830	350	0.541	108	0.168	140	0.216	185	0.287	210	0.325
27	1	26.8	1.054	564	0.873	175	0.270	226	0.349	299	0.462	338	0.524
35	1-¼	35.4	1.395	984	1.528	305	0.474	394	0.611	522	0.810	591	0.917
41	1-½	40.3	1.588	1276	1.981	395	0.614	510	0.792	676	1.050	765	1.188
53	2	51.6	2.033	2091	3.246	648	1.006	836	1.298	1108	1.720	1255	1.948

*Corresponds to 356.2(2)

TABLE 4 (continued)

Article 356—Liquidtight Flexible Nonmetallic Conduit (LFNC-A*)

Metric Designator	Trade Size	Internal Diameter		Total Area 100%		2 Wires 31%		Over 2 Wires 40%		1 Wire 53%		60%	
		mm	in.	mm^2	$in.^2$	mm^2	$in.^2$	mm^2	$in.^2$	mm^2	$in.^2$	mm^2	$in.^2$
12	⅜	12.6	0.495	125	0.192	39	0.060	50	0.077	66	0.102	75	0.115
16	½	16.0	0.630	201	0.312	62	0.097	80	0.125	107	0.165	121	0.187
21	¾	21.0	0.825	346	0.535	107	0.166	139	0.214	184	0.283	208	0.321
27	1	26.5	1.043	552	0.854	171	0.265	221	0.342	292	0.453	331	0.513
35	1-¼	35.1	1.383	968	1.502	300	0.466	387	0.601	513	0.796	581	0.901
41	1-½	40.7	1.603	1301	2.018	403	0.626	520	0.807	690	1.070	781	1.211
53	2	52.4	2.063	2157	3.343	669	1.036	863	1.337	1143	1.772	1294	2.006

*Corresponds to 356.2(1)

TABLE 4 (continued)

Article 350—Liquidtight Flexible Metal Conduit (LFMC)

Metric Designator	Trade Size	*Internal Diameter*		*Total Area 100%*		*2 Wires 31%*		*Over 2 Wires 40%*		*1 Wire 53%*		*60%*	
		mm	*in.*	*mm²*	*in.²*	*mm²*	*in.²*	*mm²*	*in.²*	*mm²*	*in.²*	*mm²*	*in.²*
12	⅜	12.5	0.494	123	0.192	38	0.059	49	0.077	65	0.102	74	0.115
16	½	16.1	0.632	204	0.314	63	0.097	81	0.125	108	0.166	122	0.188
21	¾	21.1	0.830	350	0.541	108	0.168	140	0.216	185	0.287	210	0.325
27	1	26.8	1.054	564	0.873	175	0.270	226	0.349	299	0.462	338	0.524
35	1-¼	35.4	1.395	984	1.528	305	0.474	394	0.611	522	0.810	591	0.917
41	1-½	40.3	1.588	1276	1.981	395	0.614	510	0.792	676	1.050	765	1.188
53	2	51.6	2.033	2091	3.246	648	1.006	836	1.298	1108	1.720	1255	1.948
63	2-½	63.3	2.493	3147	4.881	976	1.513	1259	1.953	1668	2.587	1888	2.929
78	3	78.4	3.085	4827	7.475	1497	2.317	1931	2.990	2559	3.962	2896	4.485
91	3-½	89.4	3.520	6277	9.731	1946	3.017	2511	3.893	3327	5.158	3766	5.839
103	4	102.1	4.020	8187	12.692	2538	3.935	3275	5.077	4339	6.727	4912	7.615
129	5	—	—	—	—	—	—	—	—	—	—	—	—
155	6	—	—	—	—	—	—	—	—	—	—	—	—

TABLE 4 (continued)

Article 344—Rigid Metal Conduit (RMC)

Metric Designator	*Trade Size*	*Internal Diameter*		*Total Area 100%*		*2 Wires 31%*		*Over 2 Wires 40%*		*1 Wire 53%*		*60%*	
		mm	*in.*	*mm²*	*in.²*	*mm²*	*in.²*	*mm²*	*in.²*	*mm²*	*in.²*	*mm²*	*in.²*
12	⅜	—	—	—	—	—	—	—	—	—	—	—	—
16	½	16.1	0.632	204	0.314	63	0.097	81	0.125	108	0.166	122	0.188
21	¾	21.2	0.836	353	0.549	109	0.170	141	0.220	187	0.291	212	0.329
27	1	27.0	1.063	573	0.887	177	0.275	229	0.355	303	0.470	344	0.532
35	1-¼	35.4	1.394	984	1.526	305	0.473	394	0.610	522	0.809	591	0.916
41	1-½	41.2	1.624	1333	2.071	413	0.642	533	0.829	707	1.098	800	1.243
53	2	52.9	2.083	2198	3.408	681	1.056	879	1.363	1165	1.806	1319	2.045
63	2-½	63.2	2.489	3137	4.866	972	1.508	1255	1.946	1663	2.579	1882	2.919
78	3	78.5	3.090	4840	7.499	1500	2.325	1936	3.000	2565	3.974	2904	4.499
91	3-½	90.7	3.570	6461	10.010	2003	3.103	2584	4.004	3424	5.305	3877	6.006
103	4	102.9	4.050	8316	12.882	2578	3.994	3326	5.153	4408	6.828	4990	7.729
129	5	128.9	5.073	13050	20.212	4045	6.266	5220	8.085	6916	10.713	7830	12.127
155	6	154.8	6.093	18821	29.158	5834	9.039	7528	11.663	9975	15.454	11292	17.495

TABLE 4 (continued)

Article 352—Rigid PVC Conduit (RNC), Schedule 80

Metric Designator	Trade Size	Internal Diameter		Total Area 100%		2 Wires 31%		Over 2 Wires 40%		1 Wire 53%		60%	
		mm	*in.*	*mm²*	*in.²*	*mm²*	*in.²*	*mm²*	*in.²*	*mm²*	*in.²*	*mm²*	*in.²*
12	⅜	—	—	—	—	—	—	—	—	—	—	—	—
16	½	13.4	0.526	141	0.217	44	0.067	56	0.087	75	0.115	85	0.130
21	¾	18.3	0.722	263	0.409	82	0.127	105	0.164	139	0.217	158	0.246
27	1	23.8	0.936	445	0.688	138	0.213	178	0.275	236	0.365	267	0.413
35	1-¼	31.9	1.255	799	1.237	248	0.383	320	0.495	424	0.656	480	0.742
41	1-½	37.5	1.476	1104	1.711	342	0.530	442	0.684	585	0.907	663	1.027
53	2	48.6	1.913	1855	2.874	575	0.891	742	1.150	983	1.523	1113	1.725
63	2-½	58.2	2.290	2660	4.119	825	1.277	1064	1.647	1410	2.183	1596	2.471
78	3	72.7	2.864	4151	6.442	1287	1.997	1660	2.577	2200	3.414	2491	3.865
91	3-½	84.5	3.326	5608	8.688	1738	2.693	2243	3.475	2972	4.605	3365	5.213
103	4	96.2	3.786	7268	11.258	2253	3.490	2907	4.503	3852	5.967	4361	6.755
129	5	121.1	4.768	11518	17.855	3571	5.535	4607	7.142	6105	9.463	6911	10.713
155	6	145.0	5.709	16513	25.598	5119	7.935	6605	10.239	8752	13.567	9908	15.359

TABLE 4 (continued)

Article 352—Rigid PVC Conduit (RNC), Schedule 40, and HDPE Conduit															
Metric Designator	*Trade Size*	*Internal Diameter*		*Total Area 100%*		*2 Wires 31%*		*Over 2 Wires 40%*		*1 Wire 53%*		*60%*			
		mm	*in.*	*mm²*	*in.²*	*mm²*	*in.²*	*mm²*	*in.²*	*mm²*	*in.²*	*mm²*	*in.²*		
12	⅜	—	—	—	—	—	—	—	—	—	—	—	—		
16	½	15.3	0.602	184	0.285	57	0.088	74	0.114	97	0.151	110	0.171		
21	¾	20.4	0.804	327	0.508	101	0.157	131	0.203	173	0.269	196	0.305		
27	1	26.1	1.029	535	0.832	166	0.258	214	0.333	284	0.441	321	0.499		
35	1-¼	34.5	1.360	935	1.453	290	0.450	374	0.581	495	0.770	561	0.872		
41	1-½	40.4	1.590	1282	1.986	397	0.616	513	0.794	679	1.052	769	1.191		
53	2	52.0	2.047	2124	3.291	658	1.020	849	1.316	1126	1.744	1274	1.975		
63	2-½	62.1	2.445	3029	4.695	939	1.455	1212	1.878	1605	2.488	1817	2.817		
78	3	77.3	3.042	4693	7.268	1455	2.253	1877	2.907	2487	3.852	2816	4.361		
91	3-½	89.4	3.521	6277	9.737	1946	3.018	2511	3.895	3327	5.161	3766	5.842		
103	4	101.5	3.998	8091	12.554	2508	3.892	3237	5.022	4288	6.654	4855	7.532		
129	5	127.4	5.016	12748	19.761	3952	6.126	5099	7.904	6756	10.473	7649	11.856		
155	6	153.2	6.031	18433	28.567	5714	8.856	7373	11.427	9770	15.141	11060	17.140		

TABLE 4 (continued)

Article 352—Type A, Rigid PVC Conduit (RNC)													
Metric Designator	Trade Size	*Internal Diameter*		*Total Area 100%*		*2 Wires 31%*		*Over 2 Wires 40%*		*1 Wire 53%*		*60%*	
		mm	*in.*	*mm²*	*in.²*	*mm²*	*in.²*	*mm²*	*in.²*	*mm²*	*in.²*	*mm²*	*in.²*
16	½	17.8	0.700	249	0.385	77	0.119	100	0.154	132	0.204	149	0.231
21	¾	23.1	0.910	419	0.650	130	0.202	168	0.260	222	0.345	251	0.390
27	1	29.8	1.175	697	1.084	216	0.336	279	0.434	370	0.575	418	0.651
35	1-¼	38.1	1.500	1140	1.767	353	0.548	456	0.707	604	0.937	684	1.060
41	1-½	43.7	1.720	1500	2.324	465	0.720	600	0.929	795	1.231	900	1.394
53	2	54.7	2.155	2350	3.647	728	1.131	940	1.459	1245	1.933	1410	2.188
63	2-½	66.9	2.635	3515	5.453	1090	1.690	1406	2.181	1863	2.890	2109	3.272
78	3	82.0	3.230	5281	8.194	1637	2.540	2112	3.278	2799	4.343	3169	4.916
91	3-½	93.7	3.690	6896	10.694	2138	3.315	2758	4.278	3655	5.668	4137	6.416
103	4	106.2	4.180	8858	13.723	2746	4.254	3543	5.489	4695	7.273	5315	8.234
129	5	—	—	—	—	—	—	—	—	—	—	—	—
155	6	—	—	—	—	—	—	—	—	—	—	—	—

TABLE 4 (continued)

Article 352—Type EB, PVC Conduit (RNC)													
Metric Designator	Trade Size	*Internal Diameter*		*Total Area 100%*		*2 Wires 31%*		*Over 2 Wires 40%*		*1 Wire 53%*		*60%*	
		mm	*in.*	*mm²*	*in.²*	*mm²*	*in.²*	*mm²*	*in.²*	*mm²*	*in.²*	*mm²*	*in.²*
16	½	—	—	—	—	—	—	—	—	—	—	—	—
21	¾	—	—	—	—	—	—	—	—	—	—	—	—
27	1	—	—	—	—	—	—	—	—	—	—	—	—
35	1-¼	—	—	—	—	—	—	—	—	—	—	—	—
41	1-½	—	—	—	—	—	—	—	—	—	—	—	—
53	2	56.4	2.221	2498	3.874	774	1.201	999	1.550	1324	2.053	1499	2.325
63	2-½	—	—	—	—	—	—	—	—	—	—	—	—
78	3	84.6	3.330	5621	8.709	1743	2.700	2248	3.484	2979	4.616	3373	5.226
91	3-½	96.6	3.804	7329	11.365	2272	3.523	2932	4.546	3884	6.023	4397	6.819
103	4	108.9	4.289	9314	14.448	2887	4.479	3726	5.779	4937	7.657	5589	8.669
129	5	135.0	5.316	14314	22.195	4437	6.881	5726	8.878	7586	11.763	8588	13.317
155	6	160.9	6.336	20333	31.530	6303	9.774	8133	12.612	10776	16.711	12200	18.918

TABLE 5 Dimensions of Insulated Conductors and Fixture Wires

Type	Size (AWG or kcmil)	Approximate Diameter		Approximate Area	
		mm	in.	mm^2	$in.^2$
Type: FFH-2, RFH-1, RFH-2, RHH*, RHW*, RHW-2*, RHH, RHW, RHW-2, SF-1, SF-2, SFF-1, SFF-2, TF, TFF, THHW, THW, THW-2, TW, XF, XFF					
RFH-2,	18	3.454	0.136	9.355	0.0145
FFH-2	16	3.759	0.148	11.10	0.0172
RHW-2, RHH,	14	4.902	0.193	18.90	0.0293
RHW	12	5.385	0.212	22.77	0.0353
RHH,	10	5.994	0.236	28.19	0.0437
RHW, RHW-2	8	8.280	0.326	53.87	0.0835
	6	9.246	0.364	67.16	0.1041
	4	10.46	0.412	86.00	0.1333
	3	11.18	0.440	98.13	0.1521
	2	11.99	0.472	112.9	0.1750
	1	14.78	0.582	171.6	0.2660
	1/0	15.80	0.622	196.1	0.3039
	2/0	16.97	0.668	226.1	0.3505
	3/0	18.29	0.720	262.7	0.4072
	4/0	19.76	0.778	306.7	0.4754
	250	22.73	0.895	405.9	0.6291
	300	24.13	0.950	457.3	0.7088
	350	25.43	1.001	507.7	0.7870
	400	26.62	1.048	556.5	0.8626
	500	28.78	1.133	650.5	1.0082
	600	31.57	1.243	782.9	1.2135
	700	33.38	1.314	874.9	1.3561
	750	34.24	1.348	920.8	1.4272
	800	35.05	1.380	965.0	1.4957
	900	36.68	1.444	1057	1.6377
	1000	38.15	1.502	1143	1.7719
	1250	43.92	1.729	1515	2.3479
	1500	47.04	1.852	1738	2.6938
	1750	49.94	1.966	1959	3.0357
	2000	52.63	2.072	2175	3.3719
SF-2, SFF-2	18	3.073	0.121	7.419	0.0115
	16	3.378	0.133	8.968	0.0139
	14	3.759	0.148	11.10	0.0172
SF-1, SFF-1	18	2.311	0.091	4.194	0.0065
RFH-1, XF, XFF	18	2.692	0.106	5.161	0.0080
TF, TFF, XF, XFF	16	2.997	0.118	7.032	0.0109
TW, XF, XFF, THHW, THW, THW-2	14	3.378	0.133	8.968	0.0139
TW, THHW, THW,	12	3.861	0.152	11.68	0.0181
THW-2	10	4.470	0.176	15.68	0.0243
	8	5.994	0.236	28.19	0.0437

TABLE 5 Dimensions of Insulated Conductors and Fixture Wires (*Continued*)

Type	Size (AWG or Kcmil)	Approximate Diameter mm	Approximate Diameter in.	Approximate Area mm^2	Approximate Area $in.^2$
RHH*, RHW*, RHW-2*	14	4.140	0.163	13.48	0.0209
	12	4.623	0.182	16.77	0.0260
Type RRH*, RHW*, RHW-2*, THHN, THHW, THW, THW-2, TFN, TFFN, THWN, THWN-2, XF, XFF					
THHW, THW, AF, XF, XFF	10	5.232	0.206	21.48	0.0333
RHH*, RHW*, RHW-2	8	6.756	0.266	35.87	0.0556
TW, THW, THHW, THW-2 RHH*, RHW* RHW-2*	6	7.722	0.304	46.84	0.0726
	4	8.941	0.352	62.77	0.0973
	3	9.652	0.380	73.16	0.1134
	2	10.46	0.412	86.00	0.1333
	1	12.50	0.492	122.6	0.1901
	1/0	13.51	0.532	143.4	0.2223
	2/0	14.68	0.578	169.3	0.2624
	3/0	16.00	0.630	201.1	0.3117
	4/0	17.48	0.688	239.9	0.3718
	250	19.43	0.765	296.5	0.4596
	300	20.83	0.820	340.7	0.5281
	350	22.12	0.871	384.4	0.5958
	400	23.32	0.918	427.0	0.6619
	500	25.48	1.003	509.7	0.7901
	600	28.27	1.113	627.7	0.9729
	700	30.07	1.184	710.3	1.1010
	750	30.94	1.218	751.7	1.1652
	800	31.75	1.250	791.7	1.2272
	900	33.38	1.314	874.9	1.3561
	1000	34.85	1.372	953.8	1.4784
	1250	39.09	1.539	1200	1.8602
	1500	42.21	1.662	1400	2.1695
	1750	45.11	1.776	1598	2.4773
	2000	47.80	1.882	1795	2.7818
TFN, TFFN	18	2.134	0.084	3.548	0.0055
	16	2.438	0.096	4.645	0.0072
THHN, THWN, THWN-2	14	2.819	0.111	6.258	0.0097
	12	3.302	0.130	8.581	0.0133
	10	4.166	0.164	13.61	0.0211
	8	5.486	0.216	23.61	0.0366
	6	6.452	0.254	32.71	0.0507
	4	8.230	0.324	53.16	0.0824
	3	8.941	0.352	62.77	0.0973
	2	9.754	0.384	74.71	0.1158
	1	11.33	0.446	100.8	0.1562
	1/0	12.34	0.486	119.7	0.1855
	2/0	13.51	0.532	143.4	0.2223

TABLE 5 Dimensions of Insulated Conductors and Fixture Wires (*Continued*)

Type	*Size (AWG or kcmil)*	*Approximate Diameter*		*Approximate Area*	
		mm	*in.*	*mm²*	*in.²*
	3/0	14.83	0.584	172.8	0.2679
	4/0	16.31	0.642	208.8	0.3237
	250	18.06	0.711	256.1	0.3970
	300	19.46	0.766	297.3	0.4608
Type: FEP, FEPB, PAF, PAFF, PF, PFA, PFAH, PFF, PGF, PGFF, PTF, PTFF, TFE, THHN, THWN, THWN-2, Z, ZF, ZFF					
THHN, THWN, THWN-2	350	20.75	0.817	338.2	0.5242
	400	21.95	0.864	378.3	0.5863
	500	24.10	0.949	456.3	0.7073
	600	26.70	1.051	559.7	0.8676
	700	28.50	1.122	637.9	0.9887
	750	29.36	1.156	677.2	1.0496
	800	30.18	1.188	715.2	1.1085
	900	31.80	1.252	794.3	1.2311
	1000	33.27	1.310	869.5	1.3478
PF, PGFF, PGF, PFF, PTF, PAF, PTFF, PAFF	18	2.184	0.086	3.742	0.0058
	16	2.489	0.098	4.839	0.0075
PF, PGFF, PGF, PFF, PTF, PAF, PTFF, PAFF, TFE, FEP, PFA, FEPB, PFAH	14	2.870	0.113	6.452	0.0100
TFE, FEP, PFA, FEPB, PFAH	12	3.353	0.132	8.839	0.0137
	10	3.962	0.156	12.32	0.0191
	8	5.232	0.206	21.48	0.0333
	6	6.198	0.244	30.19	0.0468
	4	7.417	0.292	43.23	0.0670
	3	8.128	0.320	51.87	0.0804
	2	8.941	0.352	62.77	0.0973
TFE, PFAH	1	10.72	0.422	90.26	0.1399
TFE, PFA PFAH, Z	1/0	11.73	0.462	108.1	0.1676
	2/0	12.90	0.508	130.8	0.2027
	3/0	14.22	0.560	158.9	0.2463
	4/0	15.70	0.618	193.5	0.3000
ZF, ZFF	18	1.930	0.076	2.903	0.0045
	16	2.235	0.088	3.935	0.0061
Z, ZF, ZFF	14	2.616	0.103	5.355	0.0083
Z	12	3.099	0.122	7.548	0.0117
	10	3.962	0.156	12.32	0.0191
	8	4.978	0.196	19.48	0.0302
	6	5.944	0.234	27.74	0.0430
	4	7.163	0.282	40.32	0.0625
	3	8.382	0.330	55.16	0.0855
	2	9.195	0.362	66.39	0.1029
	1	10.21	0.402	81.87	0.1269

TABLE 5 Dimensions of Insulated Conductors and Fixture Wires (*Continued*)

Type	*Size (AWG or kcmil)*	*Approximate Diameter*		*Approximate Area*	
		mm	*in.*	*mm²*	*in.²*
Type: KF-1, KF-2, KFF-1, KFF-2, XHH, XHHW, XHHW-2, ZW					
XHHW, ZW, XHHW-2, XHH	14	3.378	0.133	8.968	0.0139
	12	3.861	0.152	11.68	0.0181
	10	4.470	0.176	15.68	0.0243
	8	5.994	0.236	28.19	0.0437
	6	6.960	0.274	38.06	0.0590
	4	8.179	0.322	52.52	0.0814
	3	8.890	0.350	62.06	0.0962
	2	9.703	0.382	73.94	0.1146
XHHW, XHHW-2, XHH	1	11.23	0.442	98.97	0.1534
	1/0	12.24	0.482	117.7	0.1825
	2/0	13.41	0.528	141.3	0.2190
	3/0	14.73	0.58	170.5	0.2642
	4/0	16.21	0.638	206.3	0.3197
	250	17.91	0.705	251.9	0.3904
	300	19.30	0.76	292.6	0.4536
	350	20.60	0.811	333.3	0.5166
	400	21.79	0.858	373.0	0.5782
	500	23.95	0.943	450.6	0.6984
	600	26.75	1.053	561.9	0.8709
	700	28.55	1.124	640.2	0.9923
	750	29.41	1.158	679.5	1.0532
	800	30.23	1.190	717.5	1.1122
	900	31.85	1.254	796.8	1.2351
	1000	33.32	1.312	872.2	1.3519
	1250	37.57	1.479	1108	1.7180
	1500	40.69	1.602	1300	2.0157
	1750	43.59	1.716	1492	2.3127
	2000	46.28	1.822	1682	2.6073
KF-2, KFF-2	18	1.600	0.063	2.000	0.0031
	16	1.905	0.075	2.839	0.0044
	14	2.286	0.090	4.129	0.0064
	12	2.769	0.109	6.000	0.0093
	10	3.378	0.133	8.968	0.0139
KF-1, KFF-1	18	1.448	0.057	1.677	0.0026
	16	1.753	0.069	2.387	0.0037
	14	2.134	0.084	3.548	0.0055
	12	2.616	0.103	5.355	0.0083
	10	3.226	0.127	8.194	0.0127

*Types RHH, RHW and RHW-2 without outer covering.

TABLE 5A Compact Aluminum Building Wire Nominal Dimensions* and Areas

Size (AWG or kcmil)	Bare Conductor			Types THW and THHW				Type THHN				Type XHHW				Size (AWG or kcmil)
	Numbers of Strands	Diameter		Approximate Diameter		Approximate Area		Approximate Diameter		Approximate Area		Approximate Diameter		Approximate Area		
		mm	in.	mm	in.	mm^2	$in.^2$	mm	in.	mm^2	$in.^2$	mm	in.	mm^2	$in.^2$	
8	7	3.404	0.134	6.477	0.255	32.90	0.0510	—	—	—	—	5.690	0.224	25.42	0.0394	8
6	7	4.293	0.169	7.366	0.290	42.58	0.0660	6.096	0.240	29.16	0.0452	6.604	0.260	34.19	0.0530	6
4	7	5.410	0.213	8.509	0.335	56.84	0.0881	7.747	0.305	47.10	0.0730	7.747	0.305	47.10	0.0730	4
2	7	6.807	0.268	9.906	0.390	77.03	0.1194	9.144	0.360	65.61	0.1017	9.144	0.360	65.61	0.1017	2
1	19	7.595	0.299	11.81	0.465	109.5	0.1698	10.54	0.415	87.23	0.1352	10.54	0.415	87.23	0.1352	1
1/0	19	8.534	0.336	12.70	0.500	126.6	0.1963	11.43	0.450	102.6	0.1590	11.43	0.450	102.6	0.1590	1/0
2/0	19	9.550	0.376	13.84	0.545	150.5	0.2332	12.57	0.495	124.1	0.1924	12.45	0.490	121.6	0.1885	2/0
3/0	19	10.74	0.423	14.99	0.590	176.3	0.2733	13.72	0.540	147.7	0.2290	13.72	0.540	147.7	0.2290	3/0
4/0	19	12.07	0.475	16.38	0.645	210.8	0.3267	15.11	0.595	179.4	0.2780	14.99	0.590	176.3	0.2733	4/0

TABLE 5A (Continued)

Size (AWG or kcmil)	Bare Conductor			Types THW and THHW				Type THHN				Type XHHW				Size (AWG or kcmil)
	Number of Strands	Diameter		Approximate Diameter		Approximate Area		Approximate Diameter		Approximate Area		Approximate Diameter		Approximate Area		
		mm	in.	mm	in.	mm^2	$in.^2$	mm	in.	mm^2	$in.^2$	mm	in.	mm^2	$in.^2$	
250	37	13.21	0.520	18.42	0.725	266.3	0.4128	17.02	0.670	227.4	0.3525	16.76	0.660	220.7	0.3421	250
300	37	14.48	0.570	19.69	0.775	304.3	0.4717	18.29	0.720	262.6	0.4071	18.16	0.715	259.0	0.4015	300
350	37	15.65	0.616	20.83	0.820	340.7	0.5281	19.56	0.770	300.4	0.4656	19.30	0.760	292.6	0.4536	350
400	37	16.74	0.659	21.97	0.865	379.1	0.5876	20.70	0.815	336.5	0.5216	20.32	0.800	324.3	0.5026	400
500	37	18.69	0.736	23.88	0.940	447.7	0.6939	22.48	0.885	396.8	0.6151	22.35	0.880	392.4	0.6082	500
600	61	20.65	0.813	26.67	1.050	558.6	0.8659	25.02	0.985	491.6	0.7620	24.89	0.980	486.6	0.7542	600
700	61	22.28	0.877	28.19	1.110	624.3	0.9676	26.67	1.050	558.6	0.8659	26.67	1.050	558.6	0.8659	700
750	61	23.06	0.908	29.21	1.150	670.1	1.0386	27.31	1.075	585.5	0.9076	27.69	1.090	602.0	0.9331	750
1000	61	26.92	1.060	32.64	1.285	836.6	1.2968	31.88	1.255	798.1	1.2370	31.24	1.230	766.6	1.1882	1000

*Dimensions are from industry sources.

TABLE 8 Conductor Properties

Size (AWG or kcmil)	Area		Conductors							Direct-Current Resistance at 75°C (167°F)					
			Stranding			Overall				Copper				Aluminum	
				Diameter		Diameter		Area		Uncoated		Coated			
	mm^2	Circular Mils	Quantity	mm	in.	mm	in.	mm^2	$in.^2$	ohm/km	ohm/kFT	ohm/km	ohm/kFT	ohm/km	ohm/kFT
18	0.823	1620	1	—	—	1.02	0.040	0.823	0.001	25.5	7.77	26.5	8.08	42.0	12.8
18	0.823	1620	7	0.39	0.015	1.16	0.046	1.06	0.002	26.1	7.95	27.7	8.45	42.8	13.1
16	1.31	2580	1	—	—	1.29	0.051	1.31	0.002	16.0	4.89	16.7	5.08	26.4	8.05
16	1.31	2580	7	0.49	0.019	1.46	0.058	1.68	0.003	16.4	4.99	17.3	5.29	26.9	8.21
14	2.08	4110	1	—	—	1.63	0.064	2.08	0.003	10.1	3.07	10.4	3.19	16.6	5.06
14	2.08	4110	7	0.62	0.024	1.85	0.073	2.68	0.004	10.3	3.14	10.7	3.26	16.9	5.17
12	3.31	65.30	1	—	—	2.05	0.081	3.31	0.005	6.34	1.93	6.57	2.01	10.45	3.18
12	3.31	6530	7	0.78	0.030	2.32	0.092	4.25	0.006	6.50	1.98	6.73	2.05	10.69	3.25
10	5.261	10380	1	—	—	2.588	0.102	5.26	0.008	3.984	1.21	4.148	1.26	6.561	2.00
10	5.261	10380	7	0.98	0.038	2.95	0.116	6.76	0.011	4.070	1.24	4.226	1.29	6.679	2.04

TABLE 8 (Continued)

Size (AWG or kcmil)	Area		Conductors							Direct-Current Resistance at 75°C (167°F)					
			Stranding			Overall				Copper				Aluminum	
				Diameter		Diameter		Area		Uncoated		Coated			
	mm^2	Circular Mils	Quantity	mm	in.	mm	in.	mm^2	$in.^2$	ohm/km	ohm/kFT	ohm/km	ohm/kFT	ohm/km	ohm/kFT
8	8.367	16510	1	—	—	3.264	0.128	8.37	0.013	2.506	0.764	2.579	0.786	4.125	1.26
8	8.367	16510	7	1.23	0.049	3.71	0.146	10.76	0.017	2.551	0.778	2.653	0.809	4.204	1.28
6	13.30	26240	7	1.56	0.061	4.67	0.184	17.09	0.027	1.608	0.491	1.671	0.510	2.652	0.808
4	21.15	41740	7	1.96	0.077	5.89	0.232	27.19	0.042	1.010	0.308	1.053	0.321	1.666	0.508
3	26.67	52620	7	2.20	0.087	6.60	0.260	34.28	0.053	0.802	0.245	0.833	0.254	1.320	0.403
2	33.62	66360	7	2.47	0.097	7.42	0.292	43.23	0.067	0.634	0.194	0.661	0.201	1.045	0.319
1	42.41	83690	19	1.69	0.066	8.43	0.332	55.80	0.087	0.505	0.154	0.524	0.160	0.829	0.253
1/0	53.49	105600	19	1.89	0.074	9.45	0.372	70.41	0.109	0.399	0.122	0.415	0.127	0.660	0.201
2/0	67.43	133100	19	2.13	0.084	10.62	0.418	88.74	0.137	0.3170	0.0967	0.329	0.101	0.523	0.159
3/0	85.01	167800	19	2.39	0.094	11.94	0.470	111.9	0.173	0.2512	0.0766	0.2610	0.0797	0.413	0.126
4/0	107.2	211600	19	2.68	0.106	13.41	0.528	141.1	0.219	0.1996	0.0608	0.2050	0.0626	0.328	0.100

TABLE 8 (Continued)

Size (AWG or kcmil)	Area		Conductors							Direct-Current Resistance at 75°C (167°F)					
			Stranding			Overall				Copper				Aluminum	
				Diameter		Diameter		Area		Uncoated		Coated			
	mm²	Circular Mils	Quantity	mm	in.	mm	in.	mm²	in.²	ohm/km	ohm/kFt	ohm/km	ohm/kFT	ohm/km	ohm/kFT
250		—	37	2.09	0.082	14.61	0.575	168	0.260	0.1687	0.0515	0.1753	0.0535	0.2778	0.0847
300		—	37	2.29	0.090	16.00	0.630	201	0.312	0.1409	0.0429	0.1463	0.0446	0.2318	0.0707
350		—	37	2.47	0.097	17.30	0.681	235	0.364	0.1205	0.0367	0.1252	0.0382	0.1984	0.0605
400		—	37	2.64	0.104	18.49	0.728	268	0.416	0.1053	0.0321	0.1084	0.0331	0.1737	0.0529
500		—	37	2.95	0.116	20.65	0.813	336	0.519	0.0845	0.0258	0.0869	0.0265	0.1391	0.0424
600		—	61	2.52	0.099	22.68	0.893	404	0.626	0.0704	0.0214	0.0732	0.0223	0.1159	0.0353
700		—	61	2.72	0.107	24.49	0.964	471	0.730	0.0603	0.0184	0.0622	0.0189	0.0994	0.0303
750		—	61	2.82	0.111	25.35	0.998	505	0.782	0.0563	0.0171	0.0579	0.0176	0.0927	0.0282
800		—	61	2.91	0.114	26.16	1.030	538	0.834	0.0528	0.0161	0.0544	0.0166	0.0868	0.0265
900		—	61	3.09	0.122	27.79	1.094	606	0.940	0.0470	0.0143	0.0481	0.0147	0.0770	0.0235
1000		—	61	3.25	0.128	29.26	1.152	673	1.042	0.0423	0.0129	0.0434	0.0132	0.0695	0.0212
1250		—	91	2.98	0.117	32.74	1.289	842	1.305	0.0338	0.0103	0.0347	0.0106	0.0554	0.0169

TABLE 8 (Continued)

Size (AWG or kcmil)	Area mm²	Area Circular Mils	Stranding Quantity	Stranding Diameter mm	Stranding Diameter in.	Overall Diameter mm	Overall Diameter in.	Overall Area mm²	Overall Area in.²	Copper Uncoated ohm/km	Copper Uncoated ohm/kFT	Copper Coated ohm/km	Copper Coated ohm/kFT	Aluminum ohm/km	Aluminum ohm/kFT
1500		—	91	3.26	0.128	35.86	1.412	1011	1.566	0.02814	0.00858	0.02814	0.00883	0.0464	0.0141
1750		—	127	2.98	0.117	38.76	1.526	1180	1.829	0.02410	0.00735	0.02410	0.00756	0.0397	0.0121
2000		—	127	3.19	0.126	41.45	1.632	1349	2.092	0.02109	0.00643	0.02109	0.00662	0.0348	0.106

Conductors: Stranding, Overall. Direct-Current Resistance at 75°C (167°F).

Notes:

1. These resistance values are valid only for the parameters as given. Using conductors having coated strands, different stranding type, and, especially, other temperatures changes the resistance.
2. Formula for temperature change: $R_2 = R_1 [1 + \alpha (T_2 - 75)$, where $\alpha_{cu} = 0.00323$, $\alpha_{AL} = 0.00330$ at 75°C.
3. Conductors with compact and compressed stranding have about 9 percent and 3 percent, respectively, smaller bare conductor diameters than those shown. See Table 5A for actual compact cable dimensions.
4. The IACS conductivities used: bare copper = 100%, aluminum = 61%.
5. Class B stranding is listed as well as solid for some sizes. Its overall diameter and area is that of it circumscribing circle.

FPN: The construction information is per NEMA W68-1992 or ANSI/UL 1581–1998. The resistance is calculated per National Bureau of Standards Handbook 100, dated 1966, and Handbook 109, dated 1972.

TABLE 9 Alternating-Current Resistance and Reactance for 600-Volt Cables, 3-Phase, 60 Hz, 75°C (167°F)—Three Single Conductors in Conduit

	Ohms to Neutral per kilometer / Ohms to Neutral per 1000 Feet															
	X_L (Reactance) for All Wires		*Alternating-Current Resistance for Uncoated Copper Wires*			*Alternating-Current Resistance for Aluminum Wires*			*Effective Z at 0.85 PF for Uncoated Copper Wires*			*Effective Z at 0.85 PF for Aluminum Wires*			
Size (AWG or kcmil)	*PVC, Aluminum Conduits*	*Steel Conduit*	*PVC Conduit*	*Aluminum Conduit*	*Steel Conduit*	*PVC Conduit*	*Aluminum Conduit*	*Steel Conduit*	*PVC Conduit*	*Aluminum Conduit*	*Steel Conduit*	*PVC Conduit*	*Aluminum Conduit*	*Steel Conduit*	*Size (AWG or kcmil)*
14	0.190	0.240	10.2	10.2	10.2	—	—	—	8.9	8.9	8.9	—	—	—	14
	0.058	0.073	3.1	3.1	3.1	—	—	—	2.7	2.7	2.7	—	—	—	
12	0.177	0.223	6.6	6.6	6.6	10.5	10.5	10.5	5.6	5.6	5.6	9.2	9.2	9.2	12
	0.054	0.068	2.0	2.0	2.0	3.2	3.2	3.2	1.7	1.7	1.7	2.8	2.8	2.8	
10	0.164	0.207	3.9	3.9	3.9	6.6	6.6	6.6	3.6	3.6	3.6	5.9	5.9	5.9	10
	0.050	0.063	1.2	1.2	1.2	2.0	2.0	2.0	1.1	1.1	1.1	1.8	1.8	1.8	
8	0.171	0.213	2.56	2.56	2.56	4.3	4.3	4.3	2.26	2.26	2.30	3.6	3.6	3.6	8
	0.052	0.065	0.78	0.78	0.78	1.3	1.3	1.3	0.69	0.69	0.70	1.1	1.1	1.1	
6	0.167	0.210	1.61	1.61	1.61	2.66	2.66	2.66	1.44	1.48	1.48	2.33	2.36	2.36	6
	0.051	0.064	0.49	0.49	0.49	0.81	0.81	0.81	0.44	0.45	0.45	0.71	0.72	0.72	
4	0.157	0.197	1.02	1.02	1.02	1.67	1.67	1.67	0.95	0.95	0.98	1.51	1.51	1.51	4
	0.048	0.060	0.31	0.31	0.31	0.51	0.51	0.51	0.29	0.29	0.30	0.46	0.46	0.46	

TABLE 9 Alternating-Current Resistance and Reactance for 600-Volt Cables, 3-Phase, 60 Hz, 75°C (167°F)—Three Single Conductors in Conduit (continued)

	Ohms to Neutral per kilometer / Ohms to Neutral per 1000 Feet																
	X_L (Reactance) for All Wires		*Alternating-Current Resistance for Uncoated Copper Wires*			*Alternating-Current Resistance for Aluminum Wires*			*Effective Z at 0.85 PF for Uncoated Copper Wires*			*Effective Z at 0.85 PF for Aluminum Wires*			
Size (AWG or kcmil)	*PVC, Aluminum Conduits*	*Steel Conduit*	*PVC Conduit*	*Aluminum Conduit*	*Steel Conduit*	*PVC Conduit*	*Aluminum Conduit*	*Steel Conduit*	*PVC Conduit*	*Aluminum Conduit*	*Steel Conduit*	*PVC Conduit*	*Aluminum Conduit*	*Steel Conduit*	*Size (AWG or kcmil)*
3	0.154	0.194	0.82	0.82	0.82	1.31	1.35	1.31	0.75	0.79	0.79	1.21	1.21	1.21	3
	0.047	0.059	0.25	0.25	0.25	0.40	0.41	0.40	0.23	0.24	0.24	0.37	0.37	0.37	
2	0.148	0.187	0.62	0.66	0.66	1.05	1.05	1.05	0.62	0.62	0.66	0.98	0.98	0.98	2
	0.045	0.057	0.19	0.20	0.20	0.32	0.32	0.32	0.19	0.19	0.20	0.30	0.30	0.30	
1	0.151	0.187	0.49	0.52	0.52	0.82	0.85	0.82	0.52	0.52	0.52	0.79	0.79	0.82	1
	0.046	0.057	0.15	0.16	0.16	0.25	0.26	0.25	0.16	0.16	0.16	0.24	0.24	0.25	
1/0	0.144	0.180	0.39	0.43	0.39	0.66	0.69	0.66	0.43	0.43	0.43	0.62	0.66	0.66	1/0
	0.044	0.055	0.12	0.13	0.12	0.20	0.21	0.20	0.13	0.13	0.13	0.19	0.20	0.20	
2/0	0.141	0.177	0.33	0.33	0.33	0.52	0.52	0.52	0.36	0.36	0.36	0.52	0.52	0.52	2/0
	0.043	0.054	0.10	0.10	0.10	0.16	0.16	0.16	0.11	0.11	0.11	0.16	0.16	0.16	

TABLE 9 Alternating-Current Resistance and Reactance for 600-Volt Cables, 3-Phrase, 60 Hz, 75°C (167°F)—Three Single Conductors in Conduit (continued)

Size (AWG or kcmil)	Ohms to Neutral per kilometer / Ohms to Neutral per 1000 Feet															Size (AWG or kcmil)
	X_L (Reactance) for All Wires		Alternating-Current Resistance for Uncoated Copper Wires			Alternating-Current Resistance for Aluminum Wires			Effective Z at 0.85 PF for Uncoated Copper Wires			Effective Z at 0.85 PF for Aluminum Wires				
	PVC, Aluminum Conduits	Steel Conduit	PVC Conduit	Aluminum Conduit	Steel Conduit	PVC Conduit	Aluminum Conduit	Steel Conduit	PVC Conduit	Aluminum Conduit	Steel Conduit	PVC Conduit	Aluminum Conduit	Steel Conduit		
3/0	0.138 0.042	0.171 0.052	0.253 0.077	0.269 0.082	0.259 0.079	0.43 0.13	0.43 0.13	0.43 0.13	0.289 0.088	0.302 0.092	0.308 0.094	0.43 0.013	0.43 0.13	0.46 0.14		3/0
4/0	0.135 0.041	0.167 0.051	0.203 0.062	0.220 0.067	0.207 0.063	0.33 0.10	0.36 0.11	0.33 0.10	0.243 0.074	0.256 0.078	0.262 0.080	0.36 0.11	0.36 0.11	0.36 0.11		4/0
250	0.135 0.041	0.171 0.052	0.171 0.052	0.187 0.057	0.177 0.054	0.279 0.085	0.295 0.090	0.282 0.086	0.217 0.066	0.230 0.070	0.240 0.073	0.308 0.094	0.322 0.098	0.33 0.10		250
300	0.135 0.041	0.167 0.051	0.144 0.044	0.161 0.049	0.148 0.045	0.233 0.071	0.249 0.076	0.236 0.072	0.194 0.059	0.207 0.063	0.213 0.065	0.269 0.082	0.282 0.086	0.289 0.088		300
350	0.131 0.040	0.164 0.050	0.125 0.038	0.141 0.043	0.128 0.039	0.200 0.061	0.217 0.066	0.207 0.063	0.174 0.053	0.190 0.058	0.197 0.060	0.240 0.073	0.253 0.077	0.262 0.080		350
400	0.131 0.040	0.161 0.049	0.108 0.033	0.125 0.038	0.115 0.035	0.177 0.054	0.194 0.059	0.180 0.055	0.161 0.049	0.174 0.053	0.184 0.056	0.217 0.066	0.233 0.071	0.240 0.073		400

TABLE 9 Alternating-Current Resistance and Reactance for 600-Volt Cables, 3-Phrase, 60 Hz, 75°C (167°F)—Three Single Conductors in Conduit (continued)

Size (AWG or kcmil)	X_L (Reactance) for All Wires		Alternating-Current Resistance for Uncoated Copper Wires			Alternating-Current Resistance for Aluminum Wires			Effective Z at 0.85 PF for Uncoated Copper Wires			Effective Z at 0.85 PF for Aluminum Wires			Size (AWG or kcmil)
	Ohms to Neutral per kilometer / Ohms to Neutral per 1000 Feet														
	PVC, Aluminum Conduits	Steel Conduit	PVC Conduit	Aluminum Conduit	Steel Conduit	PVC Conduit	Aluminum Conduit	Steel Conduit	PVC Conduit	Aluminum Conduit	Steel Conduit	PVC Conduit	Aluminum Conduit	Steel Conduit	
500	0.128 0.039	0.157 0.0.48	0.089 0.027	0.105 0.032	0.095 0.029	0.141 0.043	0.157 0.048	0.148 0.045	0.141 0.043	0.157 0.048	0.164 0.050	0.187 0.057	0.200 0.061	0.210 0.064	500
600	0.128 0.039	0.157 0.048	0.075 0.023	0.092 0.028	0.0.82 0.025	0.118 0.036	0.135 0.041	0.125 0.038	0.131 0.040	0.144 0.044	0.154 0.047	0.167 0.051	0.180 0.055	0.190 0.058	600
750	0.125 0.038	0.157 0.048	0.062 0.019	0.079 0.024	0.069 0.021	0.095 0.029	0.112 0.034	0.102 0.031	0.118 0.036	0.131 0.040	0.141 0.043	0.148 0.045	0.161 0.049	0.171 0.052	750
1000	0.121 0.037	0.151 0.046	0.049 0.015	0.062 0.019	0.059 0.018	0.075 0.023	0.089 0.027	0.082 0.025	0.105 0.032	0.118 0.036	0.131 0.040	0.128 0.039	0.138 0.042	0.151 0.046	1000

Notes:

1. These values are based on the following constants: UL-Type RHH wires with Class B stranding, in cradled configuration. Wire conductivities are 100 percent IACS copper and 61 percent IACS aluminum, and aluminum conduit is 45 percent IACS. Capacitive reactance is ignored, since it is negligible at these voltages.
These resistance values are valid only at 75°C (167°F) and for the parameters as given, but are representative for 600-volt wire types operating at 60 Hz.

2. *Effective* Z is defined as $R\cos(\theta) + X\sin(\theta)$, where θ is the power factor angle of the circuit. Multiplying current by effective impedance gives a good approximation for line-to-neutral voltage drop. Effective impedance values shown in this table are valid only at 0.85 power factor.
For another circuit power factor (*PF*), effective impedance (Ze) can be calculated from R and X_L values given in this table as follows: $Ze = R \times PF + X_L \sin[\arccos(PF)]$.

TABLES 11(A) AND 11(B)

For listing purposes, Tables 11(A) and 11(B) provide the required power source limitations for Class 2 and Class 3 power sources. Table 11(A) applies for alternating-current sources, and Table 11(B) applies for direct-current sources.

The power for Class 2 and Class 3 circuits shall be either (1) inherently limited requiring no overcurrent protection, or (2) not inherently limited requiring a combination of power source and overcurrent protection. Power sources designed for interconnnection shall be listed for the purpose.

As part of the listing, the Class 2 or Class 3 power source shall be durably marked where plainly visible to indicate the class of supply and its electrical rating. A Class 2 power source not suitable for wet location use shall be so marked. Exception: limited power circuits used by listed information technology equipment.

Overcurrent devices, where required, shall be located at the point where the conductor to be protected receives its supply and shall not be interchangeable with devices of higher ratings. The overcurrent device shall be permitted as an integral part of the power source.

NOTES FOR TABLES 11(A) AND 11(B)

Note 1. V_{max}, I_{max}, and VA_{max} are determined with the current limiting impedance in the circuit (not bypassed) as follows:

V_{max}: Maximum output voltage regardless of load with rated input applied.

I_{max}: Maximum output current under any noncapacitive load, including short circuit, and with overcurrent protection bypassed if used. Where a transformer limits the output current, I_{max} limits apply after one minute of operation. Where a current-limiting impedance, listed for the purpose, or as part of a listed product, is used in combination with a nonpower-limited transformer or a stored energy source, e.g., storage battery, to limit the output current, I_{max} limits apply after five seconds.

VA_{max}: Maximum volt-ampere output after one minute of operation regardless of load and overcurrent protection bypassed if used.

Note 2. For nonsinusoidal ac, V_{max} shall be not greater than 42.4 volts peak. Where wet contact (immersion not included) is likely to occur, Class 3 wiring methods shall be used or V_{max}

shall be not greater than 15 volts for sinusoidal ac and 21.2 volts peak for nonsinusoidal ac.

Note 3. If the power source is a transformer, $(VA)_{max}$ is 350 or less when V_{max} is 15 or less.

Note 4. For dc interrupted at a rate of 10 to 200 Hz, V_{max} shall not be greater than 24.8 volts peak. Where wet contact (immersion not included) is likely to occur, Class 3 wiring methods shall be used or V_{max} shall not be greater than 30 volts for continuous dc; 12.4 volts peak for dc that is interrupted at a rate of 10 to 200 Hz.

TABLES 12(A) AND 12(B)

For listing purposes, Tables 12(A) and 12(B) provide the required power source limitations for power-limited fire alarm sources. Table 12(A) applies for alternating current sources, and Table 12(B) applies for direct current sources.

The power for power-limited fire alarm circuits shall be either (A) inherently limited requiring no overcurrent protection, or (B) not inherently limited requiring the power to be limited by a combination of power source and overcurrent protection.

As part of the listing, the PLFA power source shall be durably marked where plainly visible to indicate that it is a power-limited fire alarm power source.

The overcurrent device, where required, shall be located at the point where the conductor to be protected receives its supply and shall not be interchangeable with devices of higher ratings. The overcurrent device shall be permitted as an integral part of the power source.

NOTES FOR TABLES 12(A) AND 12(B)

Note 1: V_{max}, I_{max}, and VA_{max} are determined as follows: V_{max}: Maximum output voltage regardless of load with rated input applied.

I_{max}: Maximum output current under any noncapacitive load, including short circuit, and with overcurrent protection bypassed if used. When a transformer limits the output current, I_{max} limits apply after one minute of operation. Where a current-limiting impedance, listed for the purpose, is used in combination with a nonpower-limited transformer or a stored energy source, e.g., storage battery, to limit the output current, I_{max} limits apply after five seconds.

(*Notes continue on page 278*)

TABLE 11(A) Class 2 and Class 3 Alternating-Current Power Source Limitations

Power Source		*Inherently Limited Power Source (Overcurrent protection not required)*				*Not Inherently Limited Power Source (Overcurrent protection required)*			
		Class 2			*Class 3*	*Class 2*		*Class 3*	
Source voltage V_{max} (Volts) (see Note 1)		0 through 20*	Over 20 and through 30*	Over 30 and through 150	Over 30 and through 100	0 through 20*	Over 20 and through 30*	Over 30 and through 100	Over 100 and through 150
Power Limitations VA_{max} (Volt-Amperes) (see Note 1)		—	—	—	—	250 (see Note 3)	250	250	N.A.
Current Limitations I_{max} (Amps) (see Note 1)		8.0	8.0	0.005	$150/V_{max}$	$1000/V_{max}$	$1000/V_{max}$	$1000/V_{max}$	1.0
Maximum Overcurrent Protection (Amps)		—	—	—	—	5.0	$100/V_{max}$	$100/V_{max}$	1.0
Power Source Maximum Nameplate Rating	VA (Volt-Amps)	$5.0 \times V_{max}$	100	$0.005 \times V_{max}$	100	$5.0 \times V_{max}$	100	100	100
	Current (Amps)	5.0	$100/V_{max}$	0.005	$100/V_{max}$	5.0	$100/V_{max}$	$100/V_{max}$	$100/V_{max}$

* Voltage ranges shown are for sinusoidal ac in indoor locations or where wet contact is not likely to occur. For nonsinusoidal or wet contact conditions, see Note 2.

TABLE 11(B) Class 2 and Class 3 Direct Current Power Source Limitations

Power Source		*Inherently Limited Power Source (Overcurrent protection not required)*					*Not Inherently Limited Power Source (Overcurrent protection required)*			
		Class 2				*Class 3*	*Class 2*		*Class 3*	
Source voltage V_{max} (Volts) (see Note 1)		*0 through 20**	*Over 20 and through 30**	*Over 30 and through 60**	*Over 60 and through 150*	*Over 60 and through 100*	*0 through 20**	*Over 20 and through 60**	*Over 60 and through 100*	*Over 100 and through 150*
Power Limitations $(VA)_{max}$ (Volt-Amps) (see Note 1)		—	—	—	—	—	250 (See Note 3)	250	250	N.A.
Current Limitations I_{max} (amperes) (see Note 1)		8.0	8.0	$150/V_{max}$	0.005	$150/V_{max}$	$1000/V_{max}$	$1000/V_{max}$	$1000/V_{max}$	1.0
Maximum Overcurrent Protection (Amps)		—	—	—	—	—	5.0	$100/V_{max}$	$100/V_{max}$	1.0
Power Source Maximum Nameplate Rating	VA (Volt-Amps)	$5.0 \times V_{max}$	100	100	$0.005 \times V_{max}$	100	$5.0 \times V_{max}$	100	100	100
	Current (Amps)	5.0	$100/V_{max}$	$100/V_{max}$	0.005	$100/V_{max}$	5.0	$100/V_{max}$	$100/V_{max}$	$100/V_{max}$

*Voltage ranges shown are for continuous dc in indoor locations or where wet contact is not likely to occur. For interrupted or wet contact conditions, see Note 4.

TABLE 12(A) PLFA Alternating-Current Power Source Limitations

Power Source		*Inherently Limited Power Source (Overcurrent protection not required)*			*Not Inherently Limited Power Source (Overcurrent protection required)*		
Circuit Voltage V_{max} (Volts) (see Note 1)		*0 through 20*	*Over 20 and through 30*	*Over 30 and through 100*	*0 through 20*	*Over 20 and through 100*	*Over 100 and through 150*
Power Limitations $(VA)_{max}$ (Volt-Amps) (see Note 1)		—	—	—	250 (See Note 2)	250	N.A.
Current Limitations I_{max} (Amps) (see Note 1)		8.0	8.0	$150/V_{max}$	$1000/V_{max}$	$1000/V_{max}$	1.0
Maximum Overcurrent Protection (Amps)		—	—	—	5.0	$100/V_{MAX}$	1.0
Power Source Maximum Nameplate Ratings	VA (Volt-Amps)	$5.0 \times V_{max}$	100	100	$5.0 \times V_{max}$	100	100
	Current (Amps)	5.0	$100/V_{max}$	$100/V_{max}$	5.0	$100/V_{max}$	$100/V_{max}$

Notes found on page 273 and continued on page 278.

TABLE 12(B) PLFA Direct-Current Power Source Limitations

Power Source		*Inherently Limited Power Source (Overcurrent protection not required)*				*Not Inherently Limited Power Source (Overcurrent protection required)*		
Circuit Voltage V_{max} (Volts) (see Note 1)		*0 through 20*	*Over 20 and through 30*	*Over 30 and through 100*	*Over 100 and through 250*	*0 through 20*	*Over 20 and through 100*	*Over 100 and through 150*
Power Limitations $(VA)_{max}$ (Volt-Amps) (see Note 1)		—	—	—	—	250 (see Note 2)	250	N.A.
Current Limitations I_{max} (Amps) (see Note 1)		8.0	8.0	$150/V_{max}$	0.030	$1000/V_{max}$	$1000/V_{max}$	1.0
Maximum Overcurrent Protection (Amps)		—	—	—	—	5.0	$100/V_{max}$	1.0
Power Source Maximum Nameplate Ratings	VA (Volt-Amps)	$5.0 \times V_{max}$	100	100	$0.030 \times V_{max}$	$5.0 \times V_{max}$	100	100
	Current (Amps)	5.0	$100/V_{max}$	$100/V_{max}$	0.030	5.0	$100/V_{max}$	$100/V_{max}$

Notes found on page 273 and continued on page 278.

VA_{max}: Maximum volt-ampere output after one minute of operation regardless of load and overcurrent protection bypassed if used. Current limiting impedance shall not be bypassed when determining I_{max} and VA_{max}.

Note 2. If the power source is a transformer, $(VA)_{max}$ is 350 or less when V_{max} is 15 or less.

Annex A

Product Safety Standards

Annex B

FORMULA APPLICATION INFORMATION FOR AMPACITIES CALCULATED UNDER ENGINEERING SUPERVISION

Note: These annexes are listed for information. Refer to *NEC*® for full contents.

Annex C

CONDUIT AND TUBING FILL TABLES FOR CONDUCTORS AND FIXTURE WIRES OF THE SAME SIZE

*Where this table is used in conjunction with Tables C1 through C12, the conductors installed must be of the compact type.

Table C1 Maximum Number of Conductors or Fixture Wires in Electrical Metallic Tubing (EMT) (*Based on Table 1, Chapter 9*)

		Conductors									
	Conductor	*Metric Designator (Trade Size)*									
Type	*Size (AWG/kcmil)*	*16 (½)*	*21 (¾)*	*27 (1)*	*35 (1¼)*	*41 (1½)*	*53 (2)*	*63 (2½)*	*78 (3)*	*91 (3½)*	*103 (4)*
RHH, RHW, RHW-2	14	4	7	11	20	27	46	80	120	157	201
	12	3	6	9	17	23	38	66	100	131	167
	10	2	5	8	13	18	30	53	81	105	135
	8	1	2	4	7	9	16	28	42	55	70
	6	1	1	3	5	8	13	22	34	44	56
	4	1	1	2	4	6	10	17	26	34	44
	3	1	1	1	4	5	9	15	23	30	38
	2	1	1	1	3	4	7	13	20	26	33
	1	0	1	1	1	3	5	9	13	17	22
	1/0	0	1	1	1	2	4	7	11	15	19
	2/0	0	1	1	1	2	4	6	10	13	17
	3/0	0	0	1	1	1	3	5	8	11	14
	4/0	0	0	1	1	1	3	5	7	9	12
	250	0	0	0	1	1	1	3	5	7	9
	300	0	0	0	1	1	1	3	5	6	8
	350	0	0	0	1	1	1	3	4	6	7
	400	0	0	0	1	1	1	2	4	5	7
	500	0	0	0	0	1	1	2	3	4	6
	600	0	0	0	0	1	1	1	3	4	5
	700	0	0	0	0	0	1	1	2	3	4
	750	0	0	0	0	0	1	1	2	3	4
	800	0	0	0	0	0	1	1	2	3	4
	900	0	0	0	0	0	1	1	1	3	3
	1000	0	0	0	0	0	1	1	1	2	3
	1250	0	0	0	0	0	0	1	1	1	2
	1500	0	0	0	0	0	0	1	1	1	1
	1750	0	0	0	0	0	0	1	1	1	1
	2000	0	0	0	0	0	0	1	1	1	1
TW, THHW, THW, THW-2	14	8	15	25	43	58	96	168	254	332	424
	12	6	11	19	33	45	74	129	195	255	326
	10	5	8	14	24	33	55	96	145	190	243
	8	2	5	8	13	18	30	53	81	105	135

Table C1 *Continued*

Type	Conductor Size (AWG/kcmil)	16 (½)	21 (¾)	27 (1)	35 (1¼)	41 (1½)	53 (2)	63 (2½)	78 (3)	91 (3½)	103 (4)
		Conductors									
		Metric Designator (Trade Size)									
RHH*,	14	6	10	16	28	39	64	112	169	221	282
RHW*,	12	4	8	13	23	31	51	90	136	177	227
RHW-2*	10	3	6	10	18	24	40	70	106	138	177
	8	1	4	6	10	14	24	42	63	83	106
RHH*,	6	1	3	4	8	11	18	32	48	63	81
RHW*,	4	1	1	3	6	8	13	24	36	47	60
RHW-2*,	3	1	1	3	5	7	12	20	31	40	52
TW, THW,	2	1	1	2	4	6	10	17	26	34	44
THHW,	1	1	1	1	3	4	7	12	18	24	31
THW-2	1/0	0	1	1	2	3	6	10	16	20	26
	2/0	0	1	1	1	3	5	9	13	17	22
	3/0	0	1	1	1	2	4	7	11	15	19
	4/0	0	0	1	1	1	3	6	9	12	16
	250	0	0	1	1	1	3	5	7	10	13
	300	0	0	1	1	1	2	4	6	8	11
	350	0	0	0	1	1	1	4	6	7	10
	400	0	0	0	1	1	1	3	5	7	9
	500	0	0	0	1	1	1	3	4	6	7
	600	0	0	0	1	1	1	2	3	4	6
	700	0	0	0	0	1	1	1	3	4	5
	750	0	0	0	0	1	1	1	3	4	5
	800	0	0	0	0	1	1	1	3	3	5
	900	0	0	0	0	0	1	1	2	3	4
	1000	0	0	0	0	0	1	1	2	3	4
	1250	0	0	0	0	0	1	1	1	2	3
	1500	0	0	0	0	0	1	1	1	1	2
	1750	0	0	0	0	0	0	1	1	1	2
	2000	0	0	0	0	0	0	1	1	1	1
THHN,	14	12	22	35	61	84	138	241	364	476	608
THWN,	12	9	16	26	45	61	101	176	266	347	443
THWN-2	10	5	10	16	28	38	63	111	167	219	279
	8	3	6	9	16	22	36	64	96	126	161
	6	2	4	7	12	16	26	46	69	91	116
	4	1	2	4	7	10	16	28	43	56	71
	3	1	1	3	6	8	13	24	36	47	60
	2	1	1	3	5	7	11	20	30	40	51
	1	1	1	1	4	5	8	15	22	29	37
	1/0	1	1	1	3	4	7	12	19	25	32
	2/0	0	1	1	2	3	6	10	16	20	26
	3/0	0	1	1	1	3	5	8	13	17	22
	4/0	0	1	1	1	2	4	7	11	14	18

Table C1 *Continued*

Type	Conductor Size (AWG/kcmil)	16 (½)	21 (¾)	27 (1)	35 (1¼)	41 (1½)	53 (2)	63 (2½)	78 (3)	91 (3½)	103 (4)
		Conductors									
		Metric Designator (Trade Size)									
THHN,	250	0	0	1	1	1	3	6	9	11	15
THWN,	300	0	0	1	1	1	3	5	7	10	13
THWN-2	350	0	0	1	1	1	2	4	6	9	11
	400	0	0	0	1	1	1	4	6	8	10
	500	0	0	0	1	1	1	3	5	6	8
	600	0	0	0	1	1	1	2	4	5	7
	700	0	0	0	1	1	1	2	3	4	6
	750	0	0	0	0	1	1	1	3	4	5
	800	0	0	0	0	1	1	1	3	4	5
	900	0	0	0	0	1	1	1	3	3	4
	1000	0	0	0	0	1	1	1	2	3	4
FEP,	14	12	21	34	60	81	134	234	354	462	590
FEPB,	12	9	15	25	43	59	98	171	258	337	430
PFA,	10	6	11	18	31	42	70	122	185	241	309
PFAH,	8	3	6	10	18	24	40	70	106	138	177
TFE	6	2	4	7	12	17	28	50	75	98	126
	4	1	3	5	9	12	20	35	53	69	88
	3	1	2	4	7	10	16	29	44	57	73
	2	1	1	3	6	8	13	24	36	47	60
PFA, PFAH, TFE	1	1	1	2	4	6	9	16	25	33	42
PFAH,	1/0	1	1	1	3	5	8	14	21	27	35
TFE PFA,	2/0	0	1	1	3	4	6	11	17	22	29
PFAH,	3/0	0	1	1	2	3	5	9	14	18	24
TFE, Z	4/0	0	1	1	1	2	4	8	11	15	19
Z	14	14	25	41	72	98	161	282	426	556	711
	12	10	18	29	51	69	114	200	302	394	504
	10	6	11	18	31	42	70	122	185	241	309
	8	4	7	11	20	27	44	77	117	153	195
	6	3	5	8	14	19	31	54	82	107	137
	4	1	3	5	9	13	21	37	56	74	94
	3	1	2	4	7	9	15	27	41	54	69
	2	1	1	3	6	8	13	22	34	45	57
	1	1	1	2	4	6	10	18	28	36	46
XHH,	14	8	15	25	43	58	96	168	254	332	424
XHHW,	12	6	11	19	33	45	74	129	195	255	326
XHHW-2,	10	5	8	14	24	33	55	96	145	190	243
ZW	8	2	5	8	13	18	30	53	81	105	135
	6	1	3	6	10	14	22	39	60	78	100
	4	1	2	4	7	10	16	28	43	56	72
	3	1	1	3	6	8	14	24	36	48	61
	2	1	1	3	5	7	11	20	31	40	51

Table C1 *Continued*

		Conductors									
		Metric Designator (Trade Size)									
Type	*Conductor Size (AWG/kcmil)*	*16 (½)*	*21 (¾)*	*27 (1)*	*35 (1¼)*	*41 (1½)*	*53 (2)*	*63 (2½)*	*78 (3)*	*91 (3½)*	*103 (4)*
XHH, XHHW, XHHW-2	1	1	1	1	4	5	8	15	23	30	38
	1/0	1	1	1	3	4	7	13	19	25	32
	2/0	0	1	1	2	3	6	10	16	21	27
	3/0	0	1	1	1	3	5	9	13	17	22
	4/0	0	1	1	1	2	4	7	11	14	18
	250	0	0	1	1	1	3	6	9	12	15
	300	0	0	1	1	1	3	5	8	10	13
	350	0	0	1	1	1	2	4	7	9	11
	400	0	0	0	1	1	1	4	6	8	10
	500	0	0	0	1	1	1	3	5	6	8
	600	0	0	0	1	1	1	2	4	5	6
	700	0	0	0	0	1	1	2	3	4	6
	750	0	0	0	0	1	1	1	3	4	5
	800	0	0	0	0	1	1	1	3	4	5
	900	0	0	0	0	1	1	1	3	3	4
	1000	0	0	0	0	0	1	1	2	3	4
	1250	0	0	0	0	0	1	1	1	2	3
	1500	0	0	0	0	0	1	1	1	1	3
	1750	0	0	0	0	0	0	1	1	1	2
	2000	0	0	0	0	0	0	1	1	1	1

Fixture Wires

Type	Conductor Size (AWG/kcmil)	*Metric Designator (Trade Size)* 16 (½)	21 (¾)	27 (1)	35 (1¼)	41 (1½)	53 (2)
FFH-2, RFH-2, RFHH-3	18	8	14	24	41	56	92
	16	7	12	20	34	47	78
SF-2, SFF-2	18	10	18	30	52	71	116
	16	8	15	25	43	58	96
	14	7	12	20	34	47	78
SF-1, SFF-1	18	18	33	53	92	125	206
RFH-1, RFHH-2, TF, TFF, XF, XFF	18	14	24	39	68	92	152
RFHH-2, TF, TFF, XF, XFF	16	11	19	31	55	74	123
XF, XFF	14	8	15	25	43	58	96
TFN, TFFN	18	22	38	63	108	148	244
	16	17	29	48	83	113	186
PF, PFF, PGF, PGFF, PAF, PTF, PTFF, PAFF	18	21	36	59	103	140	231
	16	16	28	46	79	108	179
	14	12	21	34	60	81	134
ZF, ZFF, ZHF, HF, HFF	18	27	47	77	133	181	298
	16	20	35	56	98	133	220
	14	14	25	41	72	98	161
KF-2, KFF-2	18	39	69	111	193	262	433
	16	27	48	78	136	185	305
	14	19	33	54	93	127	209
	12	13	23	37	64	87	144
	10	8	15	25	43	58	96
KF-1, KFF-1	18	46	82	133	230	313	516
	16	33	57	93	161	220	362
	14	22	38	63	108	148	244
	12	14	25	41	72	98	161
	10	9	16	27	47	64	105
XF, XFF	12	4	8	13	23	31	51
	10	3	6	10	18	24	40

Note: This table is for concentric stranded conductors only. For compact stranded conductors, Table C1(A) should be used.

*Types RHH, RHW, and RHW-2 without outer covering.

Table C1(A) Maximum Number of Compact Conductors in Electrical Metallic Tubing (EMT) (*Based on Table 1, Chapter 9*)

Compact Conductors

Type	Conductor Size (AWG/kcmil)	Metric Designator (Trade Size) 16 (½)	21 (¾)	27 (1)	35 (1¼)	41 (1½)	53 (2)	63 (2½)	78 (3)	91 (3½)	103 (4)
THW, THW-2, THHW	8	2	4	6	11	16	26	46	69	90	115
	6	1	3	5	9	12	20	35	53	70	89
	4	1	2	4	6	9	15	26	40	52	67
	2	1	1	3	5	7	11	19	29	38	49
	1	1	1	1	3	4	8	13	21	27	34
	1/0	1	1	1	3	4	7	12	18	23	30
	2/0	0	1	1	2	3	5	10	15	20	25
	3/0	0	1	1	1	3	5	8	13	17	21
	4/0	0	1	1	1	2	4	7	11	14	18
	250	0	0	1	1	1	3	5	8	11	14
	300	0	0	1	1	1	3	5	7	9	12
	350	0	0	1	1	1	2	4	6	8	11
	400	0	0	0	1	1	1	4	6	8	10
	500	0	0	0	1	1	1	3	5	6	8
	600	0	0	0	1	1	1	2	4	5	7
	700	0	0	0	1	1	1	2	3	4	6
	750	0	0	0	0	1	1	1	3	4	5
	1000	0	0	0	0	1	1	1	2	3	4
THHN, THWN, THWN-2	8	—	—	—	—	—	—	—	—	—	—
	6	2	4	7	13	18	29	52	78	102	130
	4	1	3	4	8	11	18	32	48	63	81
	2	1	1	3	6	8	13	23	34	45	58
	1	1	1	2	4	6	10	17	26	34	43
	1/0	1	1	1	3	5	8	14	22	29	37
	2/0	1	1	1	3	4	7	12	18	24	30
	3/0	0	1	1	2	3	6	10	15	20	25
	4/0	0	1	1	1	3	5	8	12	16	21
	250	0	1	1	1	1	4	6	10	13	16
	300	0	0	1	1	1	3	5	8	11	14
	350	0	0	1	1	1	3	5	7	10	12
	400	0	0	1	1	1	2	4	6	9	11
	500	0	0	0	1	1	1	4	5	7	9
	600	0	0	0	1	1	1	3	4	6	7
	700	0	0	0	1	1	1	2	4	5	7
	750	0	0	0	1	1	1	2	4	5	6
	1000	0	0	0	0	1	1	1	3	3	4

Table C1(A) *Continued*

Compact Conductors

Type	Conductor Size (AWG/kcmil)	16 (½)	21 (¾)	27 (1)	35 (1¼)	41 (1½)	53 (2)	63 (2½)	78 (3)	91 (3½)	103 (4)
		Metric Designator (Trade Size)									
XHHW, XHHW-2	8	3	5	8	15	20	34	59	90	117	149
	6	1	4	6	11	15	25	44	66	87	111
	4	1	3	4	8	11	18	32	48	63	81
	2	1	1	3	6	8	13	23	34	45	58
	1	1	1	2	4	6	10	17	26	34	43
	1/0	1	1	1	3	5	8	14	22	29	37
	2/0	1	1	1	3	4	7	12	18	24	31
	3/0	0	1	1	2	3	6	10	15	20	25
	4/0	0	1	1	1	3	5	8	13	17	21
	250	0	1	1	1	2	4	7	10	13	17
	300	0	0	1	1	1	3	6	9	11	14
	350	0	0	1	1	1	3	5	8	10	13
	400	0	0	1	1	1	2	4	7	9	11
	500	0	0	0	1	1	1	4	6	7	9
	600	0	0	0	1	1	1	3	4	6	8
	700	0	0	0	1	1	1	2	4	5	7
	750	0	0	0	1	1	1	2	3	5	6
	1000	0	0	0	0	1	1	1	3	4	5

Definition: *Compact stranding* is the result of a manufacturing process where the standard conductor is compressed to the extent that the interstices (voids between strand wires) are virtually eliminated.

Table C2 Maximum Number of Conductors or Fixture Wires in Electrical Nonmetallic Tubing (ENT) (*Based on Table 1, Chapter 9*)

		Conductors					
	Conductor Size	*Metric Designator (Trade Size)*					
Type	*(AWG/kcmil)*	*16 (½)*	*21 (¾)*	*27 (1)*	*35 (1¼)*	*41 (1½)*	*53 (2)*
RHH, RHW, RHW-2	14	3	6	10	19	26	43
	12	2	5	9	16	22	36
RHH, RHW, RHW-2	10	1	4	7	13	17	29
	8	1	1	3	6	9	15
	6	1	1	3	5	7	12
	4	1	1	2	4	6	9
	3	1	1	1	3	5	8
	2	0	1	1	3	4	7
	1	0	1	1	1	3	5
	1/0	0	0	1	1	2	4
	2/0	0	0	1	1	1	3
	3/0	0	0	1	1	1	3
	4/0	0	0	1	1	1	2
	250	0	0	0	1	1	1
	300	0	0	0	1	1	1
	350	0	0	0	1	1	1
	400	0	0	0	1	1	1
	500	0	0	0	0	1	1
	600	0	0	0	0	1	1
	700	0	0	0	0	0	1
	750	0	0	0	0	0	1
	800	0	0	0	0	0	1
	900	0	0	0	0	0	1
	1000	0	0	0	0	0	1
	1250	0	0	0	0	0	0
	1500	0	0	0	0	0	0
	1750	0	0	0	0	0	0
	2000	0	0	0	0	0	0
TW, THHW, THW, THW-2	14	7	13	22	40	55	92
	12	5	10	17	31	42	71
	10	4	7	13	23	32	52
	8	1	4	7	13	17	29
RHH*, RHW*, RHW-2*	14	4	8	15	27	37	61
RHH*, RHW*, RHW-2*	12	3	7	12	21	29	49
	10	3	5	9	17	23	38
RHH*, RHW*, RH2-W*	8	1	3	5	10	14	23

Table C2 *Continued*

		Conductors					
		Metric Designator (Trade Size)					
Type	*Conductor Size (AWG/kcmil)*	*16 (½)*	*21 (¾)*	*27 (1)*	*35 (1¼)*	*41 (1½)*	*53 (2)*
RHH*, RHW*, RHW-2*, TW, THW, THHW, THW-2	6	1	2	4	7	10	17
	4	1	1	3	5	8	13
	3	1	1	2	5	7	11
	2	1	1	2	4	6	9
	1	0	1	1	3	4	6
	1/0	0	1	1	2	3	5
	2/0	0	1	1	1	3	5
	3/0	0	0	1	1	2	4
	4/0	0	0	1	1	1	3
	250	0	0	1	1	1	2
	300	0	0	0	1	1	2
	350	0	0	0	1	1	1
	400	0	0	0	1	1	1
	500	0	0	0	1	1	1
	600	0	0	0	0	1	1
	700	0	0	0	0	1	1
	750	0	0	0	0	1	1
	800	0	0	0	0	1	1
	900	0	0	0	0	0	1
	1000	0	0	0	0	0	1
	1250	0	0	0	0	0	1
	1500	0	0	0	0	0	0
	1750	0	0	0	0	0	0
	2000	0	0	0	0	0	0
THHN, THWN, THWN-2	14	10	18	32	58	80	132
	12	7	13	23	42	58	96
	10	4	8	15	26	36	60
	8	2	5	8	15	21	35
	6	1	3	6	11	15	25
	4	1	1	4	7	9	15
	3	1	1	3	5	8	13
	2	1	1	2	5	6	11
	1	1	1	1	3	5	8
	1/0	0	1	1	3	4	7
	2/0	0	1	1	2	3	5
	3/0	0	1	1	1	3	4
	4/0	0	0	1	1	2	4
	250	0	0	1	1	1	3
	300	0	0	1	1	1	2
	350	0	0	0	1	1	2
	400	0	0	0	1	1	1
	500	0	0	0	1	1	1

Table C2 *Continued*

Conductors							
		Metric Designator (Trade Size)					
Type	*Conductor Size (AWG/kcmil)*	*16 (½)*	*21 (¾)*	*27 (1)*	*35 (1¼)*	*41 (1½)*	*53 (2)*
THHN, THWN, THWN-2	600	0	0	0	1	1	1
	700	0	0	0	0	1	1
	750	0	0	0	0	1	1
	800	0	0	0	0	1	1
	900	0	0	0	0	1	1
	1000	0	0	0	0	0	1
FEP, FEPB, PFA, PFAH, TFE	14	10	18	31	56	77	128
	12	7	13	23	41	56	93
	10	5	9	16	29	40	67
	8	3	5	9	17	23	38
	6	1	4	6	12	16	27
	4	1	2	4	8	11	19
	3	1	1	4	7	9	16
	2	1	1	3	5	8	13
PFA, PFAH, TFE	1	1	1	1	4	5	9
PFA, PFAH, TFE, Z	1/0	0	1	1	3	4	7
	2/0	0	1	1	2	4	6
	3/0	0	1	1	1	3	5
	4/0	0	1	1	1	2	4
Z	14	12	22	38	68	93	154
	12	8	15	27	48	66	109
	10	5	9	16	29	40	67
	8	3	6	10	18	25	42
	6	1	4	7	13	18	30
	4	1	3	5	9	12	20
	3	1	1	3	6	9	15
	2	1	1	3	5	7	12
	1	1	1	2	4	6	10
XHH, XHHW, XHHW-2, ZW	14	7	13	22	40	55	92
	12	5	10	17	31	42	71
	10	4	7	13	23	32	52
	8	1	4	7	13	17	29
	6	1	3	5	9	13	21
	4	1	1	4	7	9	15
	3	1	1	3	6	8	13
	2	1	1	2	5	6	11

Table C2 *Continued*

		Conductors					
	Conductor Size	*Metric Designator (Trade Size)*					
Type	*(AWG/kcmil)*	*16 (½)*	*21 (¾)*	*27 (1)*	*35 (1¼)*	*41 (1½)*	*53 (2)*
XHH, XHHW, XHHW-2	1	1	1	1	3	5	8
	1/0	0	1	1	3	4	7
	2/0	0	1	1	2	3	6
	3/0	0	1	1	1	3	5
	4/0	0	0	1	1	2	4
	250	0	0	1	1	1	3
	300	0	0	1	1	1	3
	350	0	0	1	1	1	2
	400	0	0	0	1	1	1
	500	0	0	0	1	1	1
	600	0	0	0	1	1	1
	700	0	0	0	0	1	1
	750	0	0	0	0	1	1
	800	0	0	0	0	1	1
	900	0	0	0	0	1	1
	1000	0	0	0	0	0	1
	1250	0	0	0	0	0	1
	1500	0	0	0	0	0	1
	1750	0	0	0	0	0	0
	2000	0	0	0	0	0	0

Fixture Wires

Type	Conductor Size (AWG/kcmil)	Metric Designator (Trade Size) 16 (½)	21 (¾)	27 (1)	35 (1¼)	41 (1½)	53 (2)
FFH-2, RFH-2, RFHH-3	18	6	12	21	39	53	88
	16	5	10	18	32	45	74
SF-2, SFF-2	18	8	15	27	49	67	111
	16	7	13	22	40	55	92
	14	5	10	18	32	45	74
SF-1, SFF-1	18	15	28	48	86	119	197
RFH-1, RFHH-2, TF, TFF, XF, XFF	18	11	20	35	64	88	145
RFHH-2, TF, TFF, XF, XFF	16	9	16	29	51	71	117
XF, XFF	14	7	13	22	40	55	92
TFN, TFFN	18	18	33	57	102	141	233
	16	13	25	43	78	107	178
PF, PFF, PGF, PGFF, PAF, PTF, PTFF, PAFF	18	17	31	54	97	133	221
	16	13	24	42	75	103	171
	14	10	18	31	56	77	128
ZF, ZFF, ZHF, HF, HFF	18	22	40	70	125	172	285
	16	16	29	51	92	127	210
	14	12	22	38	68	93	154
KF-2, KFF-2	18	31	58	101	182	250	413
	16	22	41	71	128	176	291
	14	15	28	49	88	121	200
	12	10	19	33	60	83	138
	10	7	13	22	40	55	92
KF-1, KFF-1	18	38	69	121	217	298	493
	16	26	49	85	152	209	346
	14	18	33	57	102	141	233
	12	12	22	38	68	93	154
	10	7	14	24	44	61	101
XF, XFF	12	3	7	12	21	29	49
	10	3	5	9	17	23	38

Note: This table is for concentric stranded conductors only. For compact stranded conductors, Table C2(A) should be used.

*Types RHH, RHW, and RHW-2 without outer covering.

Table C2(A) Maximum Number of Compact Conductors in Electrical Nonmetallic Tubing (ENT) (*Based on Table 1, Chapter 9*)

Compact Conductors

		Metric Designator (Trade Size)					
Type	*Conductor Size (AWG/kcmil)*	*16 (½)*	*21 (¾)*	*27 (1)*	*35 (1¼)*	*41 (1½)*	*53 (2)*
THW, THW-2, THHW	8	1	3	6	11	15	25
	6	1	2	4	8	11	19
	4	1	1	3	6	8	14
	2	1	1	2	4	6	10
	1	0	1	1	3	4	7
	1/0	0	1	1	3	4	6
	2/0	0	1	1	2	3	5
	3/0	0	1	1	1	3	4
	4/0	0	0	1	1	2	4
	250	0	0	1	1	1	3
	300	0	0	1	1	1	2
	350	0	0	0	1	1	2
	400	0	0	0	1	1	1
	500	0	0	0	1	1	1
	600	0	0	0	1	1	1
	700	0	0	0	0	1	1
	750	0	0	0	0	1	1
	1000	0	0	0	0	0	1
THHN, THWN, THWN-2	8	—	—	—	—	—	—
	6	1	4	7	12	17	28
	4	1	2	4	7	10	17
	2	1	1	3	5	7	12
	1	1	1	2	4	5	9
	1/0	1	1	1	3	5	8
	2/0	0	1	1	3	4	6
	3/0	0	1	1	2	3	5
	4/0	0	1	1	1	2	4
	250	0	0	1	1	1	3
	300	0	0	1	1	1	3
	350	0	0	1	1	1	2
	400	0	0	0	1	1	2
	500	0	0	0	1	1	1
	600	0	0	0	1	1	1
	700	0	0	0	1	1	1

Table C2(A) *Continued*

Compact Conductors

Type	Conductor Size (AWG/kcmil)	Metric Designator (Trade Size) 16 (½)	21 (¾)	27 (1)	35 (1¼)	41 (1½)	53 (2)
THHN, THWN, THWN-2	750	0	0	0	1	1	1
	1000	0	0	0	0	1	1
XHHW, XHHW-2	8	2	4	8	14	19	32
	6	1	3	6	10	14	24
	4	1	2	4	7	10	17
	2	1	1	3	5	7	12
	1	1	1	2	4	5	9
	1/0	1	1	1	3	5	8
	2/0	0	1	1	3	4	7
	3/0	0	1	1	2	3	5
	4/0	0	1	1	1	3	4
	250	0	0	1	1	1	3
	300	0	0	1	1	1	3
	350	0	0	1	1	1	3
	400	0	0	1	1	1	2
	500	0	0	0	1	1	1
	600	0	0	0	1	1	1
	700	0	0	0	1	1	1
	750	0	0	0	1	1	1
	1000	0	0	0	0	1	1

Definition: *Compact stranding* is the result of a manufacturing process where the standard conductor is compressed to the extent that the interstices (voids between strand wires) are virtually climated.

Table C3 Maximum Number of Conductors or Fixture Wires in Flexible Metal Conduit (FMC) (*Based on Table 1, Chapter 9*)

Conductors

Type	Conductor Size (AWG/kcmil)	*Metric Designator (Trade Size)* 16 (½)	21 (¾)	27 (1)	35 (1¼)	41 (1½)	53 (2)	63 (2½)	78 (3)	91 (3½)	103 (4)
RHH, RHW, RHW-2	14	4	7	11	17	25	44	67	96	131	171
	12	3	6	9	14	21	37	55	80	109	142
RHH, RHW, RHW-2	10	3	5	7	11	17	30	45	64	88	115
	8	1	2	4	6	9	15	23	34	46	60
	6	1	1	3	5	7	12	19	27	37	48
	4	1	1	2	4	5	10	14	21	29	37
	3	1	1	1	3	5	8	13	18	25	33
	2	1	1	1	3	4	7	11	16	22	28
	1	0	1	1	1	2	5	7	10	14	19
	1/0	0	1	1	1	2	4	6	9	12	16
	2/0	0	1	1	1	1	3	5	8	11	14
	3/0	0	0	1	1	1	3	5	7	9	12
	4/0	0	0	1	1	1	2	4	6	8	10
	250	0	0	0	1	1	1	3	4	6	8
	300	0	0	0	1	1	1	2	4	5	7
	350	0	0	0	1	1	1	2	3	5	6
	400	0	0	0	0	1	1	1	3	4	6
	500	0	0	0	0	1	1	1	3	4	5
	600	0	0	0	0	1	1	1	2	3	4
	700	0	0	0	0	0	1	1	1	3	3
	750	0	0	0	0	0	1	1	1	2	3
	800	0	0	0	0	0	1	1	1	2	3
	900	0	0	0	0	0	1	1	1	2	3
	1000	0	0	0	0	0	1	1	1	1	3
	1250	0	0	0	0	0	0	1	1	1	1
	1500	0	0	0	0	0	0	1	1	1	1
	1750	0	0	0	0	0	0	1	1	1	1
	2000	0	0	0	0	0	0	0	1	1	1
TW, THHW, THW, THW-2	14	9	15	23	36	53	94	141	203	277	361
	12	7	11	18	28	41	72	108	156	212	277
	10	5	8	13	21	30	54	81	116	158	207
	8	3	5	7	11	17	30	45	64	88	115
RHH*, RHW*, RHW-2*	14	6	10	15	24	35	62	94	135	184	240
RHH*, RHW*, RHW-2*	12	5	8	12	19	28	50	75	108	148	193
	10	4	6	10	15	22	39	59	85	115	151
RHH*, RHW*, RHW-2*	8	1	4	6	9	13	23	35	51	69	90

Table C3 *Continued*

		Conductors									
		Metric Designator (Trade Size)									
Type	*Conductor Size (AWG/kcmil)*	*16 (½)*	*21 (¾)*	*27 (1)*	*35 (1¼)*	*41 (1½)*	*53 (2)*	*63 (2½)*	*78 (3)*	*91 (3½)*	*103 (4)*
RHH*, RHW*, RHW-2*, TW, THW, THHW, THW-2	6	1	3	4	7	10	18	27	39	53	69
	4	1	1	3	5	7	13	20	29	39	51
	3	1	1	3	4	6	11	17	25	34	44
	2	1	1	2	4	5	10	14	21	29	37
	1	1	1	1	2	4	7	10	15	20	26
	1/0	0	1	1	1	3	6	9	12	17	22
	2/0	0	1	1	1	3	5	7	10	14	19
	3/0	0	1	1	1	2	4	6	9	12	16
	4/0	0	0	1	1	1	3	5	7	10	13
	250	0	0	1	1	1	3	4	6	8	11
	300	0	0	1	1	1	2	3	5	7	9
	350	0	0	0	1	1	1	3	4	6	8
	400	0	0	0	1	1	1	3	4	6	7
	500	0	0	0	1	1	1	2	3	5	6
	600	0	0	0	0	1	1	1	3	4	5
	700	0	0	0	0	1	1	1	2	3	4
	750	0	0	0	0	1	1	1	2	3	4
	800	0	0	0	0	1	1	1	1	3	4
	900	0	0	0	0	0	1	1	1	3	3
	1000	0	0	0	0	0	1	1	1	2	2
	1250	0	0	0	0	0	1	1	1	1	2
	1500	0	0	0	0	0	0	1	1	1	1
	1750	0	0	0	0	0	0	1	1	1	1
	2000	0	0	0	0	0	0	1	1	1	1
THHN, THWN, THWN-2	14	13	22	33	52	76	134	202	291	396	518
	12	9	16	24	38	56	98	147	212	289	378
	10	6	10	15	24	35	62	93	134	182	238
	8	3	6	9	14	20	35	53	77	105	137
	6	2	4	6	10	14	25	38	55	76	99
	4	1	2	4	6	9	16	24	34	46	61
	3	1	1	3	5	7	13	20	29	39	51
	2	1	1	3	4	6	11	17	24	33	43
	1	1	1	1	3	4	8	12	18	24	32
	1/0	1	1	1	2	4	7	10	15	20	27
	2/0	0	1	1	1	3	6	9	12	17	22
	3/0	0	1	1	1	2	5	7	10	14	18
	4/0	0	1	1	1	1	4	6	8	12	15
	250	0	0	1	1	1	3	5	7	9	12
	300	0	0	1	1	1	3	4	6	8	11
	350	0	0	1	1	1	2	3	5	7	9
	400	0	0	0	1	1	1	3	5	6	8
	500	0	0	0	1	1	1	2	4	5	7

Table C3 *Continued*

Conductors

Type	Conductor Size (AWG/kcmil)	16 (½)	21 (¾)	27 (1)	35 (1¼)	41 (1½)	53 (2)	63 (2½)	78 (3)	91 (3½)	103 (4)
		Metric Designator (Trade Size)									
THHN, THWN, THWN-2	600	0	0	0	0	1	1	1	3	4	5
	700	0	0	0	0	1	1	1	3	4	5
	750	0	0	0	0	1	1	1	2	3	4
	800	0	0	0	0	1	1	1	2	3	4
	900	0	0	0	0	0	1	1	1	3	4
	1000	0	0	0	0	0	1	1	1	3	3
FEP, FEPB, PFA, PFAH, TFE	14	12	21	32	51	74	130	196	282	385	502
	12	9	15	24	37	54	95	143	206	281	367
	10	6	11	17	26	39	68	103	148	201	263
	8	4	6	10	15	22	39	59	85	115	151
	6	2	4	7	11	16	28	42	60	82	107
	4	1	3	5	7	11	19	29	42	57	75
	3	1	2	4	6	9	16	24	35	48	62
	2	1	1	3	5	7	13	20	29	39	51
PFA, PFAH, TFE	1	1	1	2	3	5	9	14	20	27	36
PFA, PFAH, TFE, Z	1/0	1	1	1	3	4	8	11	17	23	30
	2/0	1	1	1	2	3	6	9	14	19	24
	3/0	0	1	1	1	3	5	8	11	15	20
	4/0	0	1	1	1	2	4	6	9	13	16
Z	14	15	25	39	61	89	157	236	340	463	605
	12	11	18	28	43	63	111	168	241	329	429
	10	6	11	17	26	39	68	103	148	201	263
	8	4	7	11	17	24	43	65	93	127	166
	6	3	5	7	12	17	30	45	65	89	117
	4	1	3	5	8	12	21	31	45	61	80
	3	1	2	4	6	8	15	23	33	45	58
	2	1	1	3	5	7	12	19	27	37	49
	1	1	1	2	4	6	10	15	22	30	39
XHH, XHHW, XHHW-2, ZW	14	9	15	23	36	53	94	141	203	277	361
	12	7	11	18	28	41	72	108	156	212	277
	10	5	8	13	21	30	54	81	116	158	207
	8	3	5	7	11	17	30	45	64	88	115
	6	1	3	5	8	12	22	33	48	65	85
	4	1	2	4	6	9	16	24	34	47	61
	3	1	1	3	5	7	13	20	29	40	52
	2	1	1	3	4	6	11	17	24	33	44
XHH, XHHW, XHHW-2	1	1	1	1	3	5	8	13	18	25	32
	1/0	1	1	1	2	4	7	10	15	21	27
	2/0	0	1	1	2	3	6	9	13	17	23
	3/0	0	1	1	1	3	5	7	10	14	19
	4/0	0	1	1	1	2	4	6	9	12	15

Table C3 *Continued*

Type	Conductor Size (AWG/kcmil)	Conductors — Metric Designator (Trade Size) 16 (½)	21 (¾)	27 (1)	35 (1¼)	41 (1½)	53 (2)	63 (2½)	78 (3)	91 (3½)	103 (4)
XHH, XHHW, XHHW-2	250	0	0	1	1	1	3	5	7	10	13
	300	0	0	1	1	1	3	4	6	8	11
	350	0	0	1	1	1	2	4	5	7	9
	400	0	0	0	1	1	1	3	5	6	8
	500	0	0	0	1	1	1	3	4	5	7
	600	0	0	0	0	1	1	1	3	4	5
	700	0	0	0	0	1	1	1	3	4	5
	750	0	0	0	0	1	1	1	2	3	4
	800	0	0	0	0	1	1	1	2	3	4
	900	0	0	0	0	0	1	1	1	3	4
	1000	0	0	0	0	0	1	1	1	3	3
	1250	0	0	0	0	0	1	1	1	1	3
	1500	0	0	0	0	0	1	1	1	1	2
	1750	0	0	0	0	0	0	1	1	1	1
	2000	0	0	0	0	0	0	1	1	1	1

*Types RHH, RHW, and RHW-2 without outer covering.

Fixture Wires

Type	Conductor Size (AWG/kcmil)	Metric Designator (Trade Size) 16 (½)	21 (¾)	27 (1)	35 (1¼)	41 (1½)	53 (2)
FFH-2, RFH-2, RFHH-3	18	8	14	22	35	51	90
	16	7	12	19	29	43	76
SF-2, SFF-2	18	11	18	28	44	64	113
	16	9	15	23	36	53	94
	14	7	12	19	29	43	76
SF-1, SFF-1	18	19	32	50	78	114	201
RFH-1, RFHH-2, TF, TFF, XF, XFF	18	14	24	37	58	84	148
RFHH-2, TF, TFF, XF, XFF	16	11	19	30	47	68	120
XF, XFF	14	9	15	23	36	53	94
TFN, TFFN	18	23	38	59	93	135	237
	16	17	29	45	71	103	181
PF, PFF, PGF, PGFF, PAF, PTF, PTFF, PAFF	18	22	36	56	88	128	225
	16	17	28	43	68	99	174
	14	12	21	32	51	74	130
ZF, ZFF, ZHF, HF, HFF	18	28	47	72	113	165	290
	16	20	35	53	83	121	214
	14	15	25	39	61	89	157
KF-2, KFF-2	18	41	68	105	164	239	421
	16	28	48	74	116	168	297
	14	19	33	51	80	116	204
	12	13	23	35	55	80	140
	10	9	15	23	36	53	94
KF-1, KFF-1	18	48	82	125	196	285	503
	16	34	57	88	138	200	353
	14	23	38	59	93	135	237
	12	15	25	39	61	89	157
	10	10	16	25	40	58	103
XF, XFF	12	5	8	12	19	28	50
	10	4	6	10	15	22	39

Note: This table is for concentric stranded conductors only. For compact stranded conductors, Table C3(A) should be used.

Table C3(A) Maximum Number of Compact Conductors in Flexible Metal Conduit (FMC) (*Based on Table 1, Chapter 9*)

Compact Conductors

Type	Conductor Size (AWG/kcmil)	*Metric Designator (Trade Size)*									
		16 (½)	21 (¾)	27 (1)	35 (1¼)	41 (1½)	53 (2)	63 (2½)	78 (3)	91 (3½)	103 (4)
THW, THHW, THW-2	8	2	4	6	10	14	25	38	55	75	98
	6	1	3	5	7	11	20	29	43	58	76
	4	1	2	3	5	8	15	22	32	43	57
	2	1	1	2	4	6	11	16	23	32	42
	1	1	1	1	3	4	7	11	16	22	29
	1/0	1	1	1	2	3	6	10	14	19	25
	2/0	0	1	1	1	3	5	8	12	16	21
	3/0	0	1	1	1	2	4	7	10	14	18
	4/0	0	1	1	1	1	4	6	8	11	15
	250	0	0	1	1	1	3	4	7	9	12
	300	0	0	1	1	1	2	4	6	8	10
	350	0	0	1	1	1	2	3	5	7	0
	400	0	0	0	1	1	1	3	5	6	8
	500	0	0	0	1	1	1	3	4	5	7
	600	0	0	0	0	1	1	1	3	4	6
	700	0	0	0	0	1	1	1	3	4	5
	750	0	0	0	0	1	1	1	2	3	5
	1000	0	0	0	0	0	1	1	1	3	4
THHN, THWN, THWN-2	8	—	—	—	—	—	—	—	—	—	—
	6	3	4	7	11	16	29	43	62	85	111
	4	1	3	4	7	10	18	27	38	52	69
	2	1	1	3	5	7	13	19	28	38	49
	1	1	1	2	3	5	9	14	21	28	37
	1/0	1	1	1	3	4	8	12	17	24	31
	2/0	1	1	1	2	4	6	10	14	20	26
	3/0	0	1	1	1	3	5	8	12	17	22
	4/0	0	1	1	1	2	4	7	10	14	18
	250	0	1	1	1	1	3	5	8	11	14
	300	0	0	1	1	1	3	5	7	9	12
	350	0	0	1	1	1	3	4	6	8	10
	400	0	0	1	1	1	2	3	5	7	9
	500	0	0	0	1	1	1	3	4	6	8
	600	0	0	0	1	1	1	2	3	5	6
	700	0	0	0	0	1	1	1	3	4	6
	750	0	0	0	0	1	1	1	3	4	5
	1000	0	0	0	0	0	1	1	1	3	4
XHHW, XHHW-2	8	3	5	8	13	19	33	50	71	97	127
	6	2	4	6	9	14	24	37	53	72	95
	4	1	3	4	7	10	18	27	38	52	69
	2	1	1	3	5	7	13	19	28	38	49
	1	1	1	2	3	5	9	14	21	28	37

Table C3(A) *Continued*

Compact Conductors

Type	Conductor Size (AWG/kcmil)	Metric Designator (Trade Size) 16 (½)	21 (¾)	27 (1)	35 (1¼)	41 (1½)	53 (2)	63 (2½)	78 (3)	91 (3½)	103 (4)
XHHW, XHHW-2	1/0	1	1	1	3	4	8	12	17	24	31
	2/0	1	1	1	2	4	7	10	15	20	26
	3/0	0	1	1	1	3	5	8	12	17	22
	4/0	0	1	1	1	2	4	7	10	14	18
	250	0	1	1	1	1	4	5	8	11	14
	300	0	0	1	1	1	3	5	7	9	12
	350	0	0	1	1	1	3	4	6	8	11
	400	0	0	1	1	1	2	4	5	7	10
	500	0	0	0	1	1	1	3	4	6	8
	600	0	0	0	1	1	1	2	3	5	6
	700	0	0	0	0	1	1	1	3	4	6
	750	0	0	0	0	1	1	1	3	4	5
	1000	0	0	0	0	1	1	1	2	3	4

Definition: *Compact stranding* is the result of a manufacturing process where the standard conductor is compressed to the extent that the interstices (voids between strand wires) are virtually eliminated.

Table C4 Maximum Number of Conductors or Fixture Wires in Intermediate Metal Conduit (IMC) (*Based on Table 1, Chapter 9*)

Conductors

Type	Conductor Size (AWG/kcmil)	*Metric Designator (Trade Size)*									
		16 (½)	21 (¾)	27 (1)	35 (1¼)	41 (1½)	53 (2)	63 (2½)	78 (3)	91 (3½)	103 (4)
RHH, RHW, RHW-2	14	4	8	13	22	30	49	70	108	144	186
	12	4	6	11	18	25	41	58	89	120	154
RHH, RHW, RHW-2	10	3	5	8	15	20	33	47	72	97	124
	8	1	3	4	8	10	17	24	38	50	65
	6	1	1	3	6	8	14	19	30	40	52
	4	1	1	3	5	6	11	15	23	31	41
	3	1	1	2	4	6	9	13	21	28	36
	2	1	1	1	3	5	8	11	18	24	31
	1	0	1	1	2	3	5	7	12	16	20
	1/0	0	1	1	1	3	4	6	10	14	18
	2/0	0	1	1	1	2	4	6	9	12	15
	3/0	0	0	1	1	1	3	5	7	10	13
	4/0	0	0	1	1	1	3	4	6	9	11
	250	0	0	1	1	1	1	3	5	6	8
	300	0	0	0	1	1	1	3	4	6	7
	350	0	0	0	1	1	1	2	4	5	7
	400	0	0	0	1	1	1	2	3	5	6
	500	0	0	0	1	1	1	1	3	4	5
	600	0	0	0	0	1	1	1	2	3	4
	700	0	0	0	0	1	1	1	2	3	4
	750	0	0	0	0	1	1	1	1	3	4
	800	0	0	0	0	0	1	1	1	3	3
	900	0	0	0	0	0	1	1	1	2	3
	1000	0	0	0	0	0	1	1	1	2	3
	1250	0	0	0	0	0	1	1	1	1	2
	1500	0	0	0	0	0	0	1	1	1	1
	1750	0	0	0	0	0	0	1	1	1	1
	2000	0	0	0	0	0	0	1	1	1	1
TW, THHW, THW, THW-2	14	10	17	27	47	64	104	147	228	304	392
	12	7	13	21	36	49	80	113	175	234	301
	10	5	9	15	27	36	59	84	130	174	224
	8	3	5	8	15	20	33	47	72	97	124
RHH*, RHW*, RHW-2*	14	6	11	18	31	42	69	98	151	202	261
RHH*, RHW*, RHW-2*	12	5	9	14	25	34	56	79	122	163	209
	10	4	7	11	19	26	43	61	95	127	163
RHH*, RHW*, RHW-2*	8	2	4	7	12	16	26	37	57	76	98
RHH*, RHW*, RHW-2*	6	1	3	5	9	12	20	28	43	58	75
	4	1	2	4	6	9	15	21	32	43	56

Table C4 *Continued*

Conductors

Type	Conductor Size (AWG/kcmil)	*Metric Designator (Trade Size)* 16 (½)	21 (¾)	27 (1)	35 (1¼)	41 (1½)	53 (2)	63 (2½)	78 (3)	91 (3½)	103 (4)
TW, THW, THHW, THW-2	3	1	1	3	6	8	13	18	28	37	48
	2	1	1	3	5	6	11	15	23	31	41
	1	1	1	1	3	4	7	11	16	22	28
	1/0	1	1	1	3	4	6	9	14	19	24
	2/0	0	1	1	2	3	5	8	12	16	20
	3/0	0	1	1	1	3	4	6	10	13	17
	4/0	0	1	1	1	2	4	5	8	11	14
	250	0	0	1	1	1	3	4	7	9	12
	300	0	0	1	1	1	2	4	6	8	10
	350	0	0	1	1	1	2	3	5	7	9
	400	0	0	0	1	1	1	3	4	6	8
	500	0	0	0	1	1	1	2	4	5	7
	600	0	0	0	1	1	1	1	3	4	5
	700	0	0	0	0	1	1	1	3	4	5
	750	0	0	0	0	1	1	1	2	3	4
	800	0	0	0	0	1	1	1	2	3	4
	900	0	0	0	0	1	1	1	2	3	4
	1000	0	0	0	0	0	1	1	1	3	3
	1250	0	0	0	0	0	1	1	1	1	3
	1500	0	0	0	0	0	1	1	1	1	2
	1750	0	0	0	0	0	0	1	1	1	1
	2000	0	0	0	0	0	0	1	1	1	1
THHN, THWN, THWN-2	14	14	24	39	68	91	149	211	326	436	562
	12	10	17	29	49	67	109	154	238	318	410
	10	6	11	18	31	42	68	97	150	200	258
	8	3	6	10	18	24	39	56	86	115	149
	6	2	4	7	13	17	28	40	62	83	107
	4	1	3	4	8	10	17	25	38	51	66
	3	1	2	4	6	9	15	21	32	43	56
	2	1	1	3	5	7	12	17	27	36	47
	1	1	1	2	4	5	9	13	20	27	35
	1/0	1	1	1	3	4	8	11	17	23	29
	2/0	1	1	1	3	4	6	9	14	19	24
	3/0	0	1	1	2	3	5	7	12	16	20
	4/0	0	1	1	1	2	4	6	9	13	17
	250	0	0	1	1	1	3	5	8	10	13
	300	0	0	1	1	1	3	4	7	9	12
	350	0	0	1	1	1	2	4	6	8	10
	400	0	0	1	1	1	2	3	5	7	9
	500	0	0	0	1	1	1	3	4	6	7
	600	0	0	0	1	1	1	2	3	5	6
	700	0	0	0	1	1	1	1	3	4	5
	750	0	0	0	1	1	1	1	3	4	5
	800	0	0	0	0	1	1	1	3	4	5
	900	0	0	0	0	1	1	1	2	3	4
	1000	0	0	0	0	1	1	1	2	3	4

Table C4 *Continued*

		Conductors									
		Metric Designator (Trade Size)									
Type	*Conductor Size (AWG/kcmil)*	*16 (½)*	*21 (¾)*	*27 (1)*	*35 (1¼)*	*41 (1½)*	*53 (2)*	*63 (2½)*	*78 (3)*	*91 (3½)*	*103 (4)*
FEP, FEPB, PFA, PFAH, TFE	14	13	23	38	66	89	145	205	317	423	545
	12	10	17	28	48	65	106	150	231	309	398
	10	7	12	20	34	46	76	107	166	221	285
	8	4	7	11	19	26	43	61	95	127	163
	6	3	5	8	14	19	31	44	67	90	116
	4	1	3	5	10	13	21	30	47	63	81
	3	1	3	4	8	11	18	25	39	52	68
	2	1	2	4	6	9	15	21	32	43	56
PFA, PFAH, TFE	1	1	1	2	4	6	10	14	22	30	39
PFA, PFAH, TFE, Z	1/0	1	1	1	4	5	8	12	19	25	32
	2/0	1	1	1	3	4	7	10	15	21	27
	3/0	0	1	1	2	3	6	8	13	17	22
	4/0	0	1	1	1	3	5	7	10	14	18
Z	14	16	28	46	79	107	175	247	381	510	657
	12	11	20	32	56	76	124	175	271	362	466
	10	7	12	20	34	46	76	107	166	221	285
	8	4	7	12	21	29	48	68	105	140	180
	6	3	5	9	15	20	33	47	73	98	127
	4	1	3	6	10	14	23	33	50	67	87
	3	1	2	4	7	10	17	24	37	49	63
	2	1	1	3	6	8	14	20	30	41	53
	1	1	1	3	5	7	11	16	25	33	43
XHH, XHHW, XHHW-2, ZW	14	10	17	27	47	64	104	147	228	304	392
	12	7	13	21	36	49	80	113	175	234	301
	10	5	9	15	27	36	59	84	130	174	224
	8	3	5	8	15	20	33	47	72	97	124
	6	1	4	6	11	15	24	35	53	71	92
	4	1	3	4	8	11	18	25	39	52	67
	3	1	2	4	7	9	15	21	33	44	56
	2	1	1	3	5	7	12	18	27	37	47
XHH, XHHW, XHHW-2	1	1	1	2	4	5	9	13	20	27	35
	1/0	1	1	1	3	5	8	11	17	23	30
	2/0	1	1	1	3	4	6	9	14	19	25
	3/0	0	1	1	2	3	5	7	12	16	20
	4/0	0	1	1	1	2	4	6	10	13	17
	250	0	0	1	1	1	3	5	8	11	14
	300	0	0	1	1	1	3	4	7	9	12
	350	0	0	1	1	1	3	4	6	8	10
	400	0	0	1	1	1	2	3	5	7	9
	500	0	0	0	1	1	1	3	4	6	8

Table C4 *Continued*

Conductors

Type	Conductor Size (AWG/kcmil)	*Metric Designator (Trade Size)* 16 (½)	21 (¾)	27 (1)	35 (1¼)	41 (1½)	53 (2)	63 (2½)	78 (3)	91 (3½)	103 (4)
XHH, XHHW, XHHW-2	600	0	0	0	1	1	1	2	3	5	6
	700	0	0	0	1	1	1	1	3	4	5
	750	0	0	0	1	1	1	1	3	4	5
	800	0	0	0	0	1	1	1	3	4	5
	900	0	0	0	0	1	1	1	2	3	4
	1000	0	0	0	0	1	1	1	2	3	4
	1250	0	0	0	0	0	1	1	1	2	3
	1500	0	0	0	0	0	1	1	1	1	2
	1750	0	0	0	0	0	1	1	1	1	2
	2000	0	0	0	0	0	0	1	1	1	1

Fixture Wires

Type	Conductor Size (AWG/kcmil)	16 (½)	21 (¾)	27 (1)	35 (1¼)	41 (1 ½)	53 (2)
		Metric Designator (Trade Size)					
FHH-2, RFH-2, RFHH-3	18	9	16	26	45	61	100
	16	8	13	22	38	51	84
SF-2, SFF-2	18	12	20	33	57	77	126
	16	10	17	27	47	64	104
	14	8	13	22	38	51	84
SF-1, SFF-1	18	21	36	59	101	137	223
RFH-1, RFHH-2, TF, TFF, XF, XFF	18	15	26	43	75	101	165
RFH-2, TF, TFF, XF, XFF	16	12	21	35	60	81	133
XF, XFF	14	10	17	27	47	64	104
TFN, TFFN	18	25	42	69	119	161	264
	16	19	32	53	91	123	201
PF, PFF, PGF, PGFF, PAF, PTF, PTFF, PAFF	18	23	40	66	113	153	250
	16	18	31	51	87	118	193
	14	13	23	38	66	89	146
ZF, ZFF, ZHF, HF, HFF	18	30	52	85	146	197	322
	16	22	38	63	108	145	238
	14	16	28	46	79	107	175
KF-2, KFF-2	18	44	75	123	212	287	468
	16	31	53	87	149	202	330
	14	21	36	60	103	139	227
	12	14	25	41	70	95	156
	10	10	17	27	47	64	104
KF-1, KFF-1	18	52	90	147	253	342	558
	16	37	63	103	178	240	392
	14	25	42	69	119	161	264
	12	16	28	46	79	107	175
	10	10	18	30	52	70	114
XF, XFF	12	5	9	14	25	34	56
	10	4	7	11	19	26	43

Note: This table is for concentric stranded conductors only. For compact stranded conductors, Table C4(A) should be used.

*Types RHH, RHW, and RHW-2 without outer covering.

Table C4(A) Maximum Number of Compact Conductors in Intermediate Metal Conduit (IMC) (*Based on Table 1, Chapter 9*)

Compact Conductors

Type	Conductor Size (AWG/kcmil)	*Metric Designator (Trade Size)* 16 (½)	21 (¾)	27 (1)	35 (1¼)	41 (1½)	53 (2)	63 (2½)	78 (3)	91 (3½)	103 (4)
THW, THW-2, THHW	8	2	4	7	13	17	28	40	62	83	107
	6	1	3	6	10	13	22	31	48	64	82
	4	1	2	4	7	10	16	23	36	48	62
	2	1	1	3	5	7	12	17	26	35	45
	1	1	1	1	4	5	8	12	18	25	32
	1/0	1	1	1	3	4	7	10	16	21	27
	2/0	0	1	1	3	4	6	9	13	18	23
	3/0	0	1	1	2	3	5	7	11	15	20
	4/0	0	1	1	1	2	4	6	9	13	16
	250	0	0	1	1	1	3	5	7	10	13
	300	0	0	1	1	1	3	4	6	9	11
	350	0	0	1	1	1	2	4	6	8	10
	400	0	0	1	1	1	2	3	5	7	9
	500	0	0	0	1	1	1	3	4	6	8
	600	0	0	0	1	1	1	2	3	5	6
	700	0	0	0	1	1	1	1	3	4	5
	750	0	0	0	1	1	1	1	3	4	5
	1000	0	0	0	0	1	1	1	2	3	4
THHN, THWN, THWN-2	8	—	—	—	—	—	—	—	—	—	—
	6	3	5	8	14	19	32	45	70	95	120
	4	1	3	5	9	12	20	28	43	58	74
	2	1	1	3	6	8	14	20	31	41	53
	1	1	1	3	5	6	10	15	23	31	40
	1/0	1	1	2	4	5	9	13	20	26	34
	2/0	1	1	1	3	4	7	10	16	22	28
	3/0	0	1	1	3	4	6	9	14	18	24
	4/0	0	1	1	2	3	5	7	11	15	19
	250	0	1	1	1	2	4	6	9	12	15
	300	0	0	1	1	1	3	5	7	10	13
	350	0	0	1	1	1	3	4	7	9	11
	400	0	0	1	1	1	2	4	6	8	10
	500	0	0	1	1	1	2	3	5	7	9
	600	0	0	0	1	1	1	2	4	5	7
	700	0	0	0	1	1	1	2	3	5	6
	750	0	0	0	1	1	1	1	3	4	6
	1000	0	0	0	0	1	1	1	2	3	4

Table C4(A) *Continued*

Compact Conductors

Type	Conductor Size (AWG/kcmil)	Metric Designator (Trade Size) 16 (½)	21 (¾)	27 (1)	35 (1¼)	41 (1½)	53 (2)	63 (2½)	78 (3)	91 (3½)	103 (4)
XHHW, XHHW-2	8	3	6	9	16	22	37	52	80	107	138
	6	2	4	7	12	16	27	38	59	80	103
	4	1	3	5	9	12	20	28	43	58	74
	2	1	1	3	6	8	14	20	31	41	53
	1	1	1	3	5	6	10	15	23	31	40
	1/0	1	1	2	4	5	9	13	20	26	34
	2/0	1	1	1	3	4	7	11	17	22	29
	3/0	0	1	1	3	4	6	9	14	18	24
	4/0	0	1	1	2	3	5	7	11	15	20
	250	0	1	1	1	2	4	6	9	12	16
	300	0	0	1	1	1	3	5	8	10	13
	350	0	0	1	1	1	3	4	7	9	12
	400	0	0	1	1	1	3	4	6	8	11
	500	0	0	1	1	1	2	3	5	7	9
	600	0	0	0	1	1	1	2	4	5	7
	700	0	0	0	1	1	1	2	3	5	6
	750	0	0	0	1	1	1	1	3	4	6
	1000	0	0	0	0	1	1	1	2	3	4

Definition: *Compact stranding* is the result of a manufacturing process where the standard conductor is compressed to the extent that interstices (voids between strand wires) are virtually eliminated.

Table C5 Maximum Number of Conductors or Fixture Wires in Liquidtight Flexible Nonmetallic Conduit (Type LFNC-B*) (*Based on Table 1, Chapter 9*)

Conductors

Type	Conductor Size (AWG/kcmil)	Metric Designator (Trade Size) 12 (3/8)	16 (1/2)	21 (3/4)	27 (1)	35 (1¼)	41 (1½)	53 (2)
RHH, RHW, RHW-2	14	2	4	7	12	21	27	44
	12	1	3	6	10	17	22	36
RHH, RHW, RHW-2	10	1	3	5	8	14	18	29
	8	1	1	2	4	7	9	15
	6	1	1	1	3	6	7	12
	4	0	1	1	2	4	6	9
	3	0	1	1	1	4	5	8
	2	0	1	1	1	3	4	7
	1	0	0	1	1	1	3	5
	1/0	0	0	1	1	1	2	4
	2/0	0	0	1	1	1	1	3
	3/0	0	0	0	1	1	1	3
	4/0	0	0	0	1	1	1	2
	250	0	0	0	0	1	1	1
	300	0	0	0	0	1	1	1
	350	0	0	0	0	1	1	1
	400	0	0	0	0	1	1	1
	500	0	0	0	0	1	1	1
	600	0	0	0	0	0	1	1
	700	0	0	0	0	0	0	1
	750	0	0	0	0	0	0	1
	800	0	0	0	0	0	0	1
	900	0	0	0	0	0	0	1
	1000	0	0	0	0	0	0	1
	1250	0	0	0	0	0	0	0
	1500	0	0	0	0	0	0	0
	1750	0	0	0	0	0	0	0
	2000	0	0	0	0	0	0	0
TW, THHW, THW, THW-2	14	5	9	15	25	44	57	93
	12	4	7	12	19	33	43	71
	10	3	5	9	14	25	32	53
	8	1	3	5	8	14	18	29
RHH†, RHW†, RHW-2†	14	3	6	10	16	29	38	62
RHH†, RHW†, RHW-2†	12	3	5	8	13	23	30	50
	10	1	3	6	10	18	23	39
RHH†, RHW†, RHW-2†	8	1	1	4	6	11	14	23

Table C5 *Continued*

		Conductors						
		Metric Designator (Trade Size)						
Type	*Conductor Size (AWG/kcmil)*	*12 (⅜)*	*16 (½)*	*21 (¾)*	*27 (1)*	*35 (1¼)*	*41 (1½)*	*53 (2)*
RHH†, RHW†, RHW-2†, TW, THW	6	1	1	3	5	8	11	18
	4	1	1	1	3	6	8	13
	3	1	1	1	3	5	7	11
THHW, THW-2	2	0	1	1	2	4	6	9
	1	0	1	1	1	3	4	7
	1/0	0	0	1	1	2	3	6
	2/0	0	0	1	1	2	3	5
	3/0	0	0	1	1	1	2	4
	4/0	0	0	0	1	1	1	3
	250	0	0	0	1	1	1	3
	300	0	0	0	1	1	1	2
	350	0	0	0	0	1	1	1
	400	0	0	0	0	1	1	1
	500	0	0	0	0	1	1	1
	600	0	0	0	0	1	1	1
	700	0	0	0	0	0	1	1
	750	0	0	0	0	0	1	1
	800	0	0	0	0	0	1	1
	900	0	0	0	0	0	0	1
	1000	0	0	0	0	0	0	1
	1250	0	0	0	0	0	0	1
	1500	0	0	0	0	0	0	0
	1750	0	0	0	0	0	0	0
	2000	0	0	0	0	0	0	0
THHN, THWN, THWN-2	14	8	13	22	36	63	81	133
	12	5	9	16	26	46	59	97
	10	3	6	10	16	29	37	61
	8	1	3	6	9	16	21	35
	6	1	2	4	7	12	15	25
	4	1	1	2	4	7	9	15
	3	1	1	1	3	6	8	13
	2	1	1	1	3	5	7	11
	1	0	1	1	1	4	5	8
	1/0	0	1	1	1	3	4	7
	2/0	0	0	1	1	2	3	6
	3/0	0	0	1	1	1	3	5
	4/0	0	0	1	1	1	2	4
	250	0	0	0	1	1	1	3
	300	0	0	0	1	1	1	3
	350	0	0	0	1	1	1	2
	400	0	0	0	0	1	1	1
	500	0	0	0	0	1	1	1

Table C5 *Continued*

Type	Conductor Size (AWG/kcmil)	*Conductors* Metric Designator (Trade Size) 12 (⅜)	16 (½)	21 (¾)	27 (1)	35 (1¼)	41 (1½)	53 (2)
THHN, THWN, THWN-2	600	0	0	0	0	1	1	1
	700	0	0	0	0	1	1	1
	750	0	0	0	0	0	1	1
	800	0	0	0	0	0	1	1
	900	0	0	0	0	0	1	1
	1000	0	0	0	0	0	0	1
FEP, FEPB, PFA, PFAH, TFE	14	7	12	21	35	61	79	129
	12	5	9	15	25	44	57	94
	10	4	6	11	18	32	41	68
	8	1	3	6	10	18	23	39
	6	1	2	4	7	13	17	27
	4	1	1	3	5	9	12	19
	3	1	1	2	4	7	10	16
	2	1	1	1	3	6	8	13
PFA, PFAH, TFE	1	0	1	1	2	4	5	9
PFA, PFAH	1/0	0	1	1	1	3	4	7
TFE, Z	2/0	0	1	1	1	3	4	6
	3/0	0	0	1	1	2	3	5
	4/0	0	0	1	1	1	2	4
Z	14	9	15	26	42	73	95	156
	12	6	10	18	30	52	67	111
	10	4	6	11	18	32	41	68
	8	2	4	7	11	20	26	43
	6	1	3	5	8	14	18	30
	4	1	1	3	5	9	12	20
	3	1	1	2	4	7	9	15
	2	0	1	1	3	6	7	12
	1	0	1	1	2	5	6	10
XHH, XHHW, XHHW-2, ZW	14	5	9	15	25	44	57	93
	12	4	7	12	19	33	43	71
	10	3	5	9	14	25	32	53
	8	1	3	5	8	14	18	29
	6	1	1	3	6	10	13	22
XHH, XHHW, XHHW-2, ZW	4	1	1	2	4	7	9	16
	3	1	1	1	3	6	8	13
	2	1	1	1	3	5	7	11
XHH, XHHW, XHHW-2	1	0	1	1	1	4	5	8
	1/0	0	1	1	1	3	4	7
	2/0	0	0	1	1	2	3	6
	3/0	0	0	1	1	1	3	5
	4/0	0	0	1	1	1	2	4

Table C5 *Continued*

Conductors

Type	Conductor Size (AWG/kcmil)	Metric Designator (Trade Size) 12 (⅜)	16 (½)	21 (¾)	27 (1)	35 (1¼)	41 (1½)	53 (2)
XHH, XHHW, XHHW-2	250	0	0	0	1	1	1	3
	300	0	0	0	1	1	1	3
	350	0	0	0	1	1	1	2
	400	0	0	0	0	1	1	1
	500	0	0	0	0	1	1	1
	600	0	0	0	0	1	1	1
	700	0	0	0	0	1	1	1
	750	0	0	0	0	0	1	1
	800	0	0	0	0	0	1	1
	900	0	0	0	0	0	1	1
	1000	0	0	0	0	0	0	1
	1250	0	0	0	0	0	0	1
	1500	0	0	0	0	0	0	1
	1750	0	0	0	0	0	0	0
	2000	0	0	0	0	0	0	0
Fixture Wires								
FFH-2, RFH-2,	18	5	8	15	24	42	54	89
	16	4	7	12	20	35	46	75
SF-2, SFF-2	18	6	11	19	30	53	69	113
	16	5	9	15	25	44	57	93
	14	4	7	12	20	35	46	75
SF-1, SFF-1	18	11	19	33	53	94	122	199
RFH-1, RFHH-2, TF, TFF, XF, XFF	18	8	14	24	39	69	90	147
RFHH-2, TF, TFF, XF, XFF	16	7	11	20	32	56	72	119
XF, XFF	14	5	9	15	25	44	57	93
TFN, TFFN	18	14	23	39	63	111	144	236
	16	10	17	30	48	85	110	180
PF, PFF, PGF, PGFF, PAF, PTF, PTFF, PAFF	18	13	21	37	60	105	136	223
	16	10	16	29	46	81	105	173
	14	7	12	21	35	61	79	129
HF, HFF, ZF, ZFF, ZHF	18	17	28	48	77	136	176	288
	16	12	20	35	57	100	129	212
	14	9	15	26	42	73	95	156

Table C5 *Continued*

		Conductors						
		Metric Designator (Trade Size)						
Type	*Conductor Size (AWG/kcmil)*	*12 (⅜)*	*16 (½)*	*21 (¾)*	*27 (1)*	*35 (1¼)*	*41 (1½)*	*53 (2)*
KF-2, KFF-2	18	24	40	70	112	197	255	418
	16	17	28	49	79	139	180	295
	14	12	19	34	54	95	123	202
	12	8	13	23	37	65	85	139
	10	5	9	15	25	44	57	93
KF-1, KFF-1	18	29	48	83	134	235	304	499
	16	20	34	58	94	165	214	350
	14	14	23	39	63	111	144	236
	12	9	15	26	42	73	95	156
	10	6	10	17	27	48	62	102
XF, XFF	12	3	5	8	13	23	30	50
	10	1	3	6	10	18	23	39

Note: This table is for concentric stranded conductors only. For compact stranded conductors, Table C5(A), should be used.

*Corresponds to 356.2(2).

†Types RHH, RHW, and RHW-2 without outer covering.

Table C5(A) Maximum Number of Compact Conductors in Liquidtight Flexible Nonmetallic Conduit (Type LFNC-B*) (*Based on Table 1, Chapter 9*)

Compact Conductors

Type	Conductor Size (AWG/kcmil)	*Metric Designator (Trade Size)* 12 (3/8)	16 (1/2)	21 (3/4)	27 (1)	35 (1¼)	41 (1½)	53 (2)
THW, THW-2, THHW	8	1	2	4	7	12	15	25
	6	1	1	3	5	9	12	19
	4	1	1	2	4	7	9	14
	2	1	1	1	3	5	6	11
	1	0	1	1	1	3	4	7
	1/0	0	1	1	1	3	4	6
	2/0	0	0	1	1	2	3	5
	3/0	0	0	1	1	1	3	4
	4/0	0	0	1	1	1	2	4
	250	0	0	0	1	1	1	3
	300	0	0	0	1	1	1	2
	350	0	0	0	1	1	1	2
	400	0	0	0	0	1	1	1
	500	0	0	0	0	1	1	1
	600	0	0	0	0	1	1	1
	700	0	0	0	0	1	1	1
	750	0	0	0	0	0	1	1
	1000	0	0	0	0	0	1	1
THHN, THWN, THWN-2	8	—	—	—	—	—	—	—
	6	1	2	4	7	13	17	28
	4	1	1	3	4	8	11	17
	2	1	1	1	3	6	7	12
	1	0	1	1	2	4	6	9
	1/0	0	1	1	1	4	5	8
	2/0	0	1	1	1	3	4	6
	3/0	0	0	1	1	2	3	5
	4/0	0	0	1	1	1	3	4
	250	0	0	1	1	1	1	3
	300	0	0	0	1	1	1	3
	350	0	0	0	1	1	1	2
	400	0	0	0	1	1	1	2
	500	0	0	0	0	1	1	1
	600	0	0	0	0	1	1	1
	700	0	0	0	0	1	1	1
	750	0	0	0	0	1	1	1
	1000	0	0	0	0	0	1	1
XHHW, XHHW-2	8	1	3	5	9	15	20	33
	6	1	2	4	6	11	15	24
	4	1	1	3	4	8	11	17
	2	1	1	1	3	6	7	12
	1	0	1	1	2	4	6	9

Table C5(A) *Continued*

		Compact Conductors						
		Metric Designator (Trade Size)						
Type	*Conductor Size (AWG/kcmil)*	*12 (⅜)*	*16 (½)*	*21 (¾)*	*27 (1)*	*35 (1¼)*	*41 (1½)*	*53 (2)*
XHHW, XHHW-2	1/0	0	1	1	1	4	5	8
	2/0	0	1	1	1	3	4	7
	3/0	0	0	1	1	2	3	5
	4/0	0	0	1	1	1	3	4
	250	0	0	1	1	1	1	3
	300	0	0	0	1	1	1	3
	350	0	0	0	1	1	1	3
	400	0	0	0	1	1	1	2
	500	0	0	0	0	1	1	1
	600	0	0	0	0	1	1	1
	700	0	0	0	0	1	1	1
	750	0	0	0	0	1	1	1
	1000	0	0	0	0	0	1	1

*Corresponds to 356.2(2).

Definition: *Compact stranding* is the result of a manufacturing process where the standard conductor is compressed to the extent that the interstices (voids between strand wires) are virtually eliminated.

Table C6 Number of Conductors or Fixture Wires in Liquidtight Flexible Nonmetallic Conduit (Type LFNC-A*) (*Based on Table 1, Chapter 9*)

		Conductors						
		Metric Designator (Trade Size)						
Type	*Conductor Size (AWG/kcmil)*	*12 (3/8)*	*16 (1/2)*	*21 (3/4)*	*27 (1)*	*35 (1¼)*	*41 (1½)*	*53 (2)*
RHH,	14	2	4	7	11	20	27	45
RHW,	12	1	3	6	9	17	23	38
RHW-2	10	1	3	5	8	13	18	30
	8	1	1	2	4	7	9	16
	6	1	1	1	3	5	7	13
	4	0	1	1	2	4	6	10
	3	0	1	1	1	4	5	8
	2	0	1	1	1	3	4	7
	1	0	0	1	1	1	3	5
	1/0	0	0	1	1	1	2	4
	2/0	0	0	1	1	1	1	4
	3/0	0	0	0	1	1	1	3
	4/0	0	0	0	1	1	1	3
	250	0	0	0	0	1	1	1
	300	0	0	0	0	1	1	1
	350	0	0	0	0	1	1	1
	400	0	0	0	0	1	1	1
	500	0	0	0	0	0	1	1
	600	0	0	0	0	0	1	1
	700	0	0	0	0	0	0	1
	750	0	0	0	0	0	0	1
	800	0	0	0	0	0	0	1
	900	0	0	0	0	0	0	1
	1000	0	0	0	0	0	0	1
	1250	0	0	0	0	0	0	0
	1500	0	0	0	0	0	0	0
	1750	0	0	0	0	0	0	0
	2000	0	0	0	0	0	0	0
TW,	14	5	9	15	24	43	58	96
THHW,	12	4	7	12	19	33	44	74
THW,	10	3	5	9	14	24	33	55
THW-2	8	1	3	5	8	13	18	30
RHH[†],	14	3	6	10	16	28	38	64
RHW[†],	12	3	4	8	13	23	31	51
RHW-2[†]	10	1	3	6	10	18	24	40
	8	1	1	4	6	10	14	24
RHH[†],	6	1	1	3	4	8	11	18
RHW[†],	4	1	1	1	3	6	8	13
RHW-2[†],	3	1	1	1	3	5	7	11
TW, THW,	2	0	1	1	2	4	6	10
THHW,	1	0	1	1	1	3	4	7
THW-2								

Table C6 *Continued*

		Conductors						
		Metric Designator (Trade Size)						
Type	*Conductor Size (AWG/kcmil)*	*12 (⅜)*	*16 (½)*	*21 (¾)*	*27 (1)*	*35 (1¼)*	*41 (1½)*	*53 (2)*
RHH†, RHW†, RHW-2†, TW, THW, THHW, THW-2	1/0	0	0	1	1	2	3	6
	2/0	0	0	1	1	1	3	5
	3/0	0	0	1	1	1	2	4
	4/0	0	0	0	1	1	1	3
	250	0	0	0	1	1	1	3
	300	0	0	0	1	1	1	2
	350	0	0	0	0	1	1	1
	400	0	0	0	0	1	1	1
	500	0	0	0	0	1	1	1
	600	0	0	0	0	1	1	1
	700	0	0	0	0	0	1	1
	750	0	0	0	0	0	1	1
	800	0	0	0	0	0	1	1
	900	0	0	0	0	0	0	1
	1000	0	0	0	0	0	0	1
	1250	0	0	0	0	0	0	1
	1500	0	0	0	0	0	0	1
	1750	0	0	0	0	0	0	0
	2000	0	0	0	0	0	0	0
THHN, THWN, THWN-2	14	8	13	22	35	62	83	137
	12	5	9	16	25	45	60	100
	10	3	6	10	16	28	38	63
	8	1	3	6	9	16	22	36
	6	1	2	4	6	12	16	26
	4	1	1	2	4	7	9	16
	3	1	1	1	3	6	8	13
	2	1	1	1	3	5	7	11
	1	0	1	1	1	4	5	8
	1/0	0	1	1	1	3	4	7
	2/0	0	0	1	1	2	3	6
	3/0	0	0	1	1	1	3	5
	4/0	0	0	1	1	1	2	4
	250	0	0	0	1	1	1	3
	300	0	0	0	1	1	1	3
	350	0	0	0	1	1	1	2
	400	0	0	0	0	1	1	1
	500	0	0	0	0	1	1	1
	600	0	0	0	0	1	1	1
	700	0	0	0	0	1	1	1
	750	0	0	0	0	0	1	1
	800	0	0	0	0	0	1	1
	900	0	0	0	0	0	1	1
	1000	0	0	0	0	0	0	1

Table C6 *Continued*

		Conductors						
		Metric Designator (Trade Size)						
Type	Conductor Size (AWG/kcmil)	12 (⅜)	16 (½)	21 (¾)	27 (1)	35 (1¼)	41 (1½)	53 (2)
FEP, FEPB, PFA, PFAH, TFE	14	7	12	21	34	60	80	133
	12	5	9	15	25	44	59	97
	10	4	6	11	18	31	42	70
	8	1	3	6	10	18	24	40
	6	1	2	4	7	13	17	28
	4	1	1	3	5	9	12	20
	3	1	1	2	4	7	10	16
	2	1	1	1	3	6	8	13
PFA, PFAH, TFE	1	0	1	1	2	4	5	9
PFA, PFAH, TFE, Z	1/0	0	1	1	1	3	5	8
	2/0	0	1	1	1	3	4	6
	3/0	0	0	1	1	2	3	5
	4/0	0	0	1	1	1	2	4
Z	14	9	15	25	41	72	97	161
	12	6	10	18	29	51	69	114
	10	4	6	11	18	31	42	70
	8	2	4	7	11	20	26	44
	6	1	3	5	8	14	18	31
	4	1	1	3	5	9	13	21
	3	1	1	2	4	7	9	15
	2	1	1	1	3	6	8	13
	1	1	1	1	2	4	6	10
XHH, XHHW, XHHW-2, ZW	14	5	9	15	24	43	58	96
	12	4	7	12	19	33	44	74
	10	3	5	9	14	24	33	55
	8	1	3	5	8	13	18	30
	6	1	1	3	5	10	13	22
	4	1	1	2	4	7	10	16
	3	1	1	1	3	6	8	14
	2	1	1	1	3	5	7	11
XHH, XHHW, XHHW-2	1	0	1	1	1	4	5	8
	1/0	0	1	1	1	3	4	7
	2/0	0	0	1	1	2	3	6
	3/0	0	0	1	1	1	3	5
	4/0	0	0	1	1	1	2	4
	250	0	0	0	1	1	1	3
	300	0	0	0	1	1	1	3
	350	0	0	0	1	1	1	2
	400	0	0	0	0	1	1	1
	500	0	0	0	0	1	1	1

Table C6 *Continued*

Type	Conductor Size (AWG/kcmil)	12 (⅜)	16 (½)	21 (¾)	27 (1)	35 (1¼)	41 (1½)	53 (2)
		Conductors						
		Metric Designator (Trade Size)						
XHH, XHHW, XHHW-2	600	0	0	0	0	1	1	1
	700	0	0	0	0	1	1	1
	750	0	0	0	0	0	1	1
	800	0	0	0	0	0	1	1
	900	0	0	0	0	0	1	1
	1000	0	0	0	0	0	0	1
	1250	0	0	0	0	0	0	1
	1500	0	0	0	0	0	0	1
	1750	0	0	0	0	0	0	0
	2000	0	0	0	0	0	0	0
		Fixture Wires						
FFH-2, RFH-2,	18	5	8	14	23	41	55	92
	16	4	7	12	20	35	47	77
RFHH-3	16	4	7	12	20	35	47	77
SF-2, SFF-2	18	6	11	18	29	52	70	116
	16	5	9	15	24	43	58	96
	14	4	7	12	20	35	47	77

Table C6 *Continued*

		Conductors						
		Metric Designator (Trade Size)						
Type	*Conductor Size (AWG/kcmil)*	*12 (⅜)*	*16 (½)*	*21 (¾)*	*27 (1)*	*35 (1¼)*	*41 (1½)*	*53 (2)*
SF-1, SFF-1	18	12	19	33	52	92	124	205
RFH-1, RFHH-2, TF, TFF, XF, XFF	18	8	14	24	39	68	91	152
RFHH-2, TF, TFF, XF, XFF	16	7	11	19	31	55	74	122
XF, XFF	14	5	9	15	24	43	58	96
TFN, TFFN	18	14	22	39	62	109	146	243
	16	10	17	29	47	83	112	185
PF, PFF, PGF, PGFF, PAF, PTF, PTFF, PAFF	18	13	21	37	59	103	139	230
	16	10	16	28	45	80	107	178
	14	7	12	21	34	60	80	133
HF, HFF, ZF, ZFF, ZHF	18	17	27	47	76	133	179	297
	16	12	20	35	56	98	132	219
	14	9	15	25	41	72	97	161
KF-2, KFF-2	18	25	40	69	110	193	260	431
	16	17	28	48	77	136	183	303
	14	12	19	33	53	94	126	209
	12	8	13	23	36	64	86	143
	10	5	9	15	24	43	58	96
KF-1, KFF-1	18	29	48	82	131	231	310	514
	16	21	33	57	92	162	218	361
	14	14	22	39	62	109	146	243
	12	9	15	25	41	72	97	161
	10	6	10	17	27	47	63	105
XF, XFF	12	3	4	8	13	23	31	51
	10	1	3	6	10	18	24	40

Note: This table is for concentric stranded conductors only. For compact stranded conductors, Table C6(A) should be used.

*Corresponds to 356.2(1).

†Types RHH, RHW, and RHW-2 without outer covering.

Table C6(A) Maximum Number of Compact Conductors in Liquidtight Flexible Nonmetallic Conduit (Type LFNC-A*) (*Based on Table 1, Chapter 9*)

		Compact Conductors						
		Metric Designator (Trade Size)						
Type	*Conductor Size (AWG/kcmil)*	*12 (⅜)*	*16 (½)*	*21 (¾)*	*27 (1)*	*35 (1¼)*	*41 (1½)*	*53 (2)*
THW, THW-2, THHW	8	1	2	4	6	11	16	26
	6	1	1	3	5	9	12	20
	4	1	1	2	4	7	9	15
	2	1	1	1	3	5	6	11
	1	0	1	1	1	3	4	8
	1/0	0	1	1	1	3	4	7
	2/0	0	0	1	1	2	3	5
	3/0	0	0	1	1	1	3	5
	4/0	0	0	1	1	1	2	4
	250	0	0	0	1	1	1	3
	300	0	0	0	1	1	1	3
	350	0	0	0	1	1	1	2
	400	0	0	0	0	1	1	1
	500	0	0	0	0	1	1	1
	600	0	0	0	0	1	1	1
	700	0	0	0	0	1	1	1
	750	0	0	0	0	0	1	1
	1000	0	0	0	0	0	1	1
THHN, THWN, THWN-2	8	—	—	—	—	—	—	—
	6	1	2	4	7	13	18	29
	4	1	1	3	4	8	11	18
	2	1	1	1	3	6	8	13
	1	0	1	1	2	4	6	10
	1/0	0	1	1	1	3	5	8
	2/0	0	1	1	1	3	4	7
	3/0	0	0	1	1	2	3	6
	4/0	0	0	1	1	1	3	5
	250	0	0	1	1	1	1	3
	300	0	0	0	1	1	1	3
	350	0	0	0	1	1	1	3
	400	0	0	0	1	1	1	2
	500	0	0	0	0	1	1	1
	600	0	0	0	0	1	1	1
	700	0	0	0	0	1	1	1
	750	0	0	0	0	1	1	1
	1000	0	0	0	0	0	1	1

Table C6(A) *Continued*

		Compact Conductors						
		Metric Designator (Trade Size)						
Type	*Conductor Size (AWG/kcmil)*	*12 (⅜)*	*16 (½)*	*21 (¾)*	*27 (1)*	*35 (1¼)*	*41 (1½)*	*53 (2)*
XHHW, XHHW-2	8	1	3	5	8	15	20	34
	6	1	2	4	6	11	15	25
	4	1	1	3	4	8	11	18
	2	1	1	1	3	6	8	13
	1	0	1	1	2	4	6	10
	1/0	0	1	1	1	3	5	8
	2/0	0	1	1	1	3	4	7
	3/0	0	0	1	1	2	3	6
	4/0	0	0	1	1	1	3	5
	250	0	0	1	1	1	2	4
	300	0	0	0	1	1	1	3
	350	0	0	0	1	1	1	3
	400	0	0	0	1	1	1	2
	500	0	0	0	0	1	1	1
	600	0	0	0	0	1	1	1
	700	0	0	0	0	1	1	1
	750	0	0	0	0	1	1	1
	1000	0	0	0	0	0	1	1

*Corresponds to 356.2(1).

Definition: *Compact stranding* is the result of a manufacturing process where the standard conductor is compressed to the extent that the interstices (voids between strand wires) are virtually eliminated.

Table C7 Maximum Number of Conductors or Fixture Wires in Liquidtight Flexible Metal Conduit (LFML) (*Based on Table 1, Chapter 9*)

Conductors

Type	Conductor Size (AWG/kcmil)	*Metric Designator (Trade Size)* 16 (½)	21 (¾)	27 (1)	35 (1¼)	41 (1½)	53 (2)	63 (2½)	78 (3)	91 (3½)	103 (4)
RHH, RHW, RHW-2	14	4	7	12	21	27	44	66	102	133	173
	12	3	6	10	17	22	36	55	84	110	144
	10	3	5	8	14	18	29	44	68	89	116
	8	1	2	4	7	9	15	23	36	46	61
	6	1	1	3	6	7	12	18	28	37	48
	4	1	1	2	4	6	9	14	22	29	38
	3	1	1	1	4	5	8	13	19	25	33
	2	1	1	1	3	4	7	11	17	22	29
	1	0	1	1	1	3	5	7	11	14	19
	1/0	0	1	1	1	2	4	6	10	13	16
	2/0	0	1	1	1	1	3	5	8	11	14
	3/0	0	0	1	1	1	3	4	7	9	12
	4/0	0	0	1	1	1	2	4	6	8	10
	250	0	0	0	1	1	1	3	4	6	8
	300	0	0	0	1	1	1	2	4	5	7
	350	0	0	0	1	1	1	2	3	5	6
	400	0	0	0	1	1	1	1	3	4	6
	500	0	0	0	1	1	1	1	3	4	5
	600	0	0	0	0	1	1	1	2	3	4
	700	0	0	0	0	0	1	1	1	3	3
	750	0	0	0	0	0	1	1	1	2	3
	800	0	0	0	0	0	1	1	1	2	3
	900	0	0	0	0	0	1	1	1	2	3
	1000	0	0	0	0	0	1	1	1	1	3
	1250	0	0	0	0	0	0	1	1	1	1
	1500	0	0	0	0	0	0	1	1	1	1
	1750	0	0	0	0	0	0	1	1	1	1
	2000	0	0	0	0	0	0	0	1	1	1
TW, THHW, THW, THW-2	14	9	15	25	44	57	93	140	215	280	365
	12	7	12	19	33	43	71	108	165	215	280
	10	5	9	14	25	32	53	80	123	160	209
	8	3	5	8	14	18	29	44	68	89	116
RHH*, RHW*, RHW-2*	14	6	10	16	29	38	62	93	143	186	243
RHH*, RHW*, RHW-2*	12	5	8	13	23	30	50	75	115	149	195
	10	3	6	10	18	23	39	58	89	117	152
	8	1	4	6	11	14	23	35	53	70	91

Table C7 *Continued*

		Conductors									
		Metric Designator (Trade Size)									
Type	*Conductor Size (AWG/kcmil)*	*16 (½)*	*21 (¾)*	*27 (1)*	*35 (1¼)*	*41 (1½)*	*53 (2)*	*63 (2½)*	*78 (3)*	*91 (3½)*	*103 (4)*
RHH*,	6	1	3	5	8	11	18	27	41	53	70
RHW*,	4	1	1	3	6	8	13	20	30	40	52
RHW-2*,	3	1	1	3	5	7	11	17	26	34	44
TW,	2	1	1	2	4	6	9	14	22	29	38
THW,	1	1	1	1	3	4	7	10	15	20	26
THHW,	1/0	0	1	1	2	3	6	8	13	17	23
THW-2	2/0	0	1	1	2	3	5	7	11	15	19
	3/0	0	1	1	1	2	4	6	9	12	16
	4/0	0	0	1	1	1	3	5	8	10	13
	250	0	0	1	1	1	3	4	6	8	11
	300	0	0	1	1	1	2	3	5	7	9
	350	0	0	0	1	1	1	3	5	6	8
	400	0	0	0	1	1	1	3	4	6	7
	500	0	0	0	1	1	1	2	3	5	6
	600	0	0	0	1	1	1	1	3	4	5
	700	0	0	0	0	1	1	1	2	3	4
	750	0	0	0	0	1	1	1	2	3	4
	800	0	0	0	0	1	1	1	2	3	4
	900	0	0	0	0	0	1	1	1	3	3
	1000	0	0	0	0	0	1	1	1	2	3
	1250	0	0	0	0	0	1	1	1	1	2
	1500	0	0	0	0	0	0	1	1	1	2
	1750	0	0	0	0	0	0	1	1	1	1
	2000	0	0	0	0	0	0	1	1	1	1
THHN,	14	13	22	36	63	81	133	201	308	401	523
THWN,	12	9	16	26	46	59	97	146	225	292	381
THWN-2	10	6	10	16	29	37	61	92	141	184	240
	8	3	6	9	16	21	35	53	81	106	138
	6	2	4	7	12	15	25	38	59	76	100
	4	1	2	4	7	9	15	23	36	47	61
	3	1	1	3	6	8	13	20	30	40	52
	2	1	1	3	5	7	11	17	26	33	44
	1	1	1	1	4	5	8	12	19	25	22
	1/0	1	1	1	3	4	7	10	16	21	27
	2/0	0	1	1	2	3	6	8	13	17	23
	3/0	0	1	1	1	3	5	7	11	14	19
	4/0	0	1	1	1	2	4	6	9	12	15
	250	0	0	1	1	1	3	5	7	10	12
	300	0	0	1	1	1	3	4	6	8	11
	350	0	0	1	1	1	2	3	5	7	9
	400	0	0	0	1	1	1	3	5	6	8
	500	0	0	0	1	1	1	2	4	5	7
	600	0	0	0	1	1	1	1	3	4	6
	700	0	0	0	1	1	1	1	3	4	5
	750	0	0	0	0	1	1	1	3	3	5
	800	0	0	0	0	1	1	1	2	3	4
	900	0	0	0	0	1	1	1	2	3	4
	1000	0	0	0	0	0	1	1	1	3	3

Table C7 *Continued*

		Conductors									
	Conductor	*Metric Designator (Trade Size)*									
Type	*Size (AWG/kcmil)*	*16 (½)*	*21 (¾)*	*27 (1)*	*35 (1¼)*	*41 (1½)*	*53 (2)*	*63 (2½)*	*78 (3)*	*91 (3½)*	*103 (4)*
FEP,	14	12	21	35	61	79	129	195	299	389	507
FEPB,	12	9	15	25	44	57	94	142	218	284	370
PFA,	10	6	11	18	32	41	68	102	156	203	266
PFAH,	8	3	6	10	18	23	39	58	89	117	152
TFE	6	2	4	7	13	17	27	41	64	83	108
	4	1	3	5	9	12	19	29	44	58	75
	3	1	2	4	7	10	16	24	37	48	63
	2	1	1	3	6	8	13	20	30	40	52
PFA, PFAH, TFE	1	1	1	2	4	5	9	14	21	28	36
PFA,	1/0	1	1	1	3	4	7	11	18	23	30
PFAH,	2/0	1	1	1	3	4	6	9	14	19	25
TFE, Z	3/0	0	1	1	2	3	5	8	12	16	20
	4/0	0	1	1	1	2	4	6	10	13	17
Z	14	20	26	42	73	95	156	235	360	469	611
	12	14	18	30	52	67	111	167	255	332	434
	10	8	11	18	32	41	68	102	156	203	266
	8	5	7	11	20	26	43	64	99	129	168
	6	4	5	8	14	18	30	45	69	90	118
	4	2	3	5	9	12	20	31	48	62	81
	3	2	2	4	7	9	15	23	35	45	59
	2	1	1	3	6	7	12	19	29	38	49
	1	1	1	2	5	6	10	15	23	30	40
XHH,	14	9	15	25	44	57	93	140	215	280	365
XHHW,	12	7	12	19	33	43	71	108	165	215	280
XHHW-2,	10	5	9	14	25	32	53	80	123	160	209
ZW	8	3	5	8	14	18	29	44	68	89	116
	6	1	3	6	10	13	22	33	50	66	86
	4	1	2	4	7	9	16	24	36	48	62
	3	1	1	3	6	8	13	20	31	40	52
	2	1	1	3	5	7	11	17	26	34	44
XHH,	1	1	1	1	4	5	8	12	19	25	33
XHHW,	1/0	1	1	1	3	4	7	10	16	21	28
XHHW-2	2/0	0	1	1	2	3	6	9	13	17	23
	3/0	0	1	1	1	3	5	7	11	14	19
	4/0	0	1	1	1	2	4	6	9	12	16
	250	0	0	1	1	1	3	5	7	10	13
	300	0	0	1	1	1	3	4	6	8	11
	350	0	0	1	1	1	2	3	5	7	10
	400	0	0	0	1	1	1	3	5	6	8
	500	0	0	0	1	1	1	2	4	5	7

Table C7 *Continued*

Conductors

Type	Conductor Size (AWG/kcmil)	16 (½)	21 (¾)	27 (1)	35 (1¼)	41 (1½)	53 (2)	63 (2½)	78 (3)	91 (3½)	103 (4)
		Metric Designator (Trade Size)									
XHH, XHHW, XHHW-2	600	0	0	0	1	1	1	1	3	4	6
	700	0	0	0	1	1	1	1	3	4	5
	750	0	0	0	0	1	1	1	3	3	5
	800	0	0	0	0	1	1	1	2	3	4
	900	0	0	0	0	1	1	1	2	3	4
	1000	0	0	0	0	0	1	1	1	3	3
	1250	0	0	0	0	0	1	1	1	1	3
	1500	0	0	0	0	0	1	1	1	1	2
	1750	0	0	0	0	0	0	1	1	1	2
	2000	0	0	0	0	0	0	1	1	1	2

*Types RHH, RHW, and RHW-2 without outer covering.

Fixture Wires

Type	Conductor Size (AWG/kcmil)	16 (½)	21 (¾)	27 (1)	35 (1¼)	41 (1½)	53 (2)
		Metric Designator (Trade Size)					
FFH-2, RFH-2, RFHH-3	18	8	15	24	42	54	89
	16	7	12	20	35	46	75
SF-2, SFF-2	18	11	19	30	53	69	113
	16	9	15	25	44	57	93
	14	7	12	20	35	46	75
SF-1, SFF-1	18	19	33	53	94	122	199
RFH-1, RFHH-2, TF, TFF, XF, XFF	18	14	24	39	69	90	147
RFHH-2, TF, TFF, XF, XFF	16	11	20	32	56	72	119
XF, XFF	14	9	15	25	44	57	93
TFN, TFFN	18	23	39	63	111	144	236
	16	17	30	48	85	110	180
PF, PFF, PGF, PGFF, PAF, PTF, PTFF, PAFF	18	21	37	60	105	136	223
	16	16	29	46	81	105	173
	14	12	21	35	61	79	129

Fixture Wires *Continued*

Type	Conductor Size (AWG/kcmil)	Metric Designator (Trade Size) 16 (½)	21 (¾)	27 (1)	35 (1¼)	41 (1½)	53 (2)
HF, HFF, ZF, ZFF, ZHF	18	28	48	77	136	176	288
	16	20	35	57	100	129	212
	14	15	26	42	73	95	156
KF-2, KFF-2	18	40	70	112	197	255	418
	16	28	49	79	139	180	295
	14	19	34	54	95	123	202
	12	13	23	37	65	85	139
	10	9	15	25	44	57	93
KF-1, KFF-1	18	48	83	134	235	304	499
	16	34	58	94	165	214	350
	14	23	39	63	111	144	236
	12	15	26	42	73	95	156
	10	10	17	27	48	62	102
XF, XFF	12	5	8	13	23	30	50
	10	3	6	10	18	23	39

Note: This table is for concentric stranded conductors only. For compact stranded conductors, Table C7(A) should be used.

Table C7(A) Maximum Number of Compact Conductors in Liquidtight Flexible Metal Conduit (LFMC) (*Based on Table 1, Chapter 9*)

Compact Conductors

Type	Conductor Size (AWG/kcmil)	*Metric Designator (Trade Size)* 12 (3/8)	16 (1/2)	21 (3/4)	27 (1)	35 (1¼)	41 (1½)	53 (2)	63 (2½)	78 (3)	91 (3½)	1 (
THW, THW-2, THHW	8	1	2	4	7	12	15	25	38	58	76	
	6	1	1	3	5	9	12	19	29	45	59	
	4	1	1	2	4	7	9	14	22	34	44	
	2	1	1	1	3	5	6	11	16	25	32	
	1	0	1	1	1	3	4	7	11	17	23	
	1/0	0	1	1	1	3	4	6	10	15	20	
	2/0	0	0	1	1	2	3	5	8	13	16	
	3/0	0	0	1	1	1	3	4	7	11	14	
	4/0	0	0	1	1	1	2	4	6	9	12	
	250	0	0	0	1	1	1	3	4	7	9	
	300	0	0	0	1	1	1	2	4	6	8	
	350	0	0	0	1	1	1	2	3	5	7	
	400	0	0	0	0	1	1	1	3	5	6	
	500	0	0	0	0	1	1	1	3	4	5	
	600	0	0	0	0	1	1	1	1	3	4	
	700	0	0	0	0	1	1	1	1	3	4	
	750	0	0	0	0	0	1	1	1	3	3	
	1000	0	0	0	0	0	1	1	1	1	3	
THHN, THWN, THWN-2	8	—	—	—	—	—	—	—	—	—	—	
	6	1	2	4	7	13	17	28	43	66	86	1
	4	1	1	3	4	8	11	17	26	41	53	
	2	1	1	1	3	6	7	12	19	29	38	
	1	0	1	1	2	4	6	9	14	22	28	
	1/0	0	1	1	1	4	5	8	12	19	24	
	2/0	0	1	1	1	3	4	6	10	15	20	
	3/0	0	0	1	1	2	3	5	8	13	17	
	4/0	0	0	1	1	1	3	4	7	10	14	
	250	0	0	1	1	1	1	3	5	8	11	
	300	0	0	0	1	1	1	3	4	7	9	
	350	0	0	0	1	1	1	2	4	6	8	
	400	0	0	0	1	1	1	2	3	5	7	
	500	0	0	0	0	1	1	1	3	5	6	
	600	0	0	0	0	1	1	1	2	4	5	
	700	0	0	0	0	1	1	1	1	3	4	
	750	0	0	0	0	1	1	1	1	3	4	
	1000	0	0	0	0	0	1	1	1	2	3	
XHHW, XHHW-2	8	1	3	5	9	15	20	33	49	76	98	1
	6	1	2	4	6	11	15	24	37	56	73	
	4	1	1	3	4	8	11	17	26	41	53	
	2	1	1	1	3	6	7	12	19	29	38	
	1	0	1	1	2	4	6	9	14	22	28	

Table C7(A) *Continued*

Compact Conductors

Type	Conductor Size (AWG/kcmil)	*12* (⅜)	*16* (½)	*21* (¾)	*27* (1)	*35* (1¼)	*41* (1½)	*53* (2)	*63* (2½)	*78* (3)	*91* (3½)	*103* (4)
		Metric Designator (Trade Size)										
XHHW, XHHW-2	1/0	0	1	1	1	4	5	8	12	19	24	32
	2/0	0	1	1	1	3	4	7	10	16	20	27
	3/0	0	0	1	1	2	3	5	8	13	17	22
	4/0	0	0	1	1	1	3	4	7	11	14	18
	250	0	0	1	1	1	1	3	5	8	11	15
	300	0	0	0	1	1	1	3	5	7	9	12
	350	0	0	0	1	1	1	3	4	6	8	11
	400	0	0	0	1	1	1	2	4	6	7	10
	500	0	0	0	0	1	1	1	3	5	6	8
	600	0	0	0	0	1	1	1	2	4	5	6
	700	0	0	0	0	1	1	1	1	3	4	6
	750	0	0	0	0	1	1	1	1	3	4	5
	1000	0	0	0	0	0	1	1	1	2	3	4

Definition: *Compact stranding* is the result of a manufacturing process where the standard conductor is compressed to the extent that the interstices (voids between strand wires) are virtually eliminated.

Table C8 Maximum Number of Conductors or Fixture Wires in Rigid Metal Conduit (RMC) (*Based on Table 1, Chapter 9*)

		Conductors											
		Metric Designator (Trade Size)											
Type	*Conductor Size (AWG/kcmil)*	*16 (½)*	*21 (¾)*	*27 (1)*	*35 (1¼)*	*41 (1½)*	*53 (2)*	*63 (2½)*	*78 (3)*	*91 (3½)*	*103 (4)*	*129 (5)*	*155 (6)*
RHH, RHW, RHW-2	14	4	7	12	21	28	46	66	102	136	176	276	398
	12	3	6	10	17	23	38	55	85	113	146	229	330
	10	3	5	8	14	19	31	44	68	91	118	185	267
	8	1	2	4	7	10	16	23	36	48	61	97	139
	6	1	1	3	6	8	13	18	29	38	49	77	112
	4	1	1	2	4	6	10	14	22	30	38	60	87
	3	1	1	2	4	5	9	12	19	26	34	53	76
	2	1	1	1	3	4	7	11	17	23	29	46	66
	1	0	1	1	1	3	5	7	11	15	19	30	44
	1/0	0	1	1	1	2	4	6	10	13	17	26	38
	2/0	0	1	1	1	2	4	5	8	11	14	23	33
	3/0	0	0	1	1	1	3	4	7	10	12	30	28
	4/0	0	0	1	1	1	3	4	6	8	11	17	24
	250	0	0	0	1	1	1	3	4	6	8	13	18
	300	0	0	0	1	1	1	2	4	5	7	11	16
	350	0	0	0	1	1	1	2	4	5	6	10	15
	400	0	0	0	1	1	1	1	3	4	6	9	13
	500	0	0	0	1	1	1	1	3	4	5	8	11

	600	0	0	0	0	1	1	1	2	3	4	6	9
	700	0	0	0	0	1	1	1	1	3	4	6	8
	750	0	0	0	0	0	1	1	1	3	3	5	8
	800	0	0	0	0	0	1	1	1	2	3	5	7
	900	0	0	0	0	0	1	1	1	2	3	5	7
	1000	0	0	0	0	0	1	1	1	1	3	4	6
	1250	0	0	0	0	0	0	1	1	1	1	3	5
	1500	0	0	0	0	0	0	1	1	1	1	3	4
	1750	0	0	0	0	0	0	1	1	1	1	2	4
	2000	0	0	0	0	0	0	0	1	1	1	2	3
TW,	14	9	15	25	44	59	98	140	216	288	370	581	839
THHW,	12	7	12	19	33	45	75	107	165	221	284	446	644
THW,	10	5	9	14	25	34	56	80	123	164	212	332	480
THW-2	8	3	5	8	14	19	31	44	68	91	118	185	267
RHH*, RHW*, RHW-2*	14	6	10	17	29	39	65	93	143	191	246	387	558
RHH*, RHW*, RHW-2*	12	5	8	13	23	32	52	75	115	154	198	311	448
	10	3	6	10	18	25	41	58	90	120	154	242	350
RHH*, RHW*, RHW-2*	8	1	4	6	11	15	24	35	54	72	92	145	209

Table C8 *Continued*

		Conductors											
		Metric Designator (Trade Size)											
Type	*Conductor Size (AWG/kcmil)*	*16 (½)*	*21 (¾)*	*27 (1)*	*35 (1¼)*	*41 (1½)*	*53 (2)*	*63 (2½)*	*78 (3)*	*91 (3½)*	*103 (4)*	*129 (5)*	*155 (6)*
RHH*,	6	1	3	5	8	11	18	27	41	55	71	111	160
RHW*,	4	1	1	3	6	8	14	20	31	41	53	83	120
RHW-2*,	3	1	1	3	5	7	12	17	26	35	45	71	103
TW,	2	1	1	2	4	6	10	14	22	30	38	60	87
THW,	1	1	1	1	3	4	7	10	15	21	27	42	61
THHW,	1/0	0	1	1	2	3	6	8	13	18	23	36	52
THW-2	2/0	0	1	1	2	3	5	7	11	15	19	31	44
	3/0	0	1	1	1	2	4	6	9	13	16	26	37
	4/0	0	0	1	1	1	3	5	8	10	14	21	31
	250	0	0	1	1	1	3	4	6	8	11	17	25
	300	0	0	1	1	1	2	3	5	7	9	15	22
	350	0	0	0	1	1	1	3	5	6	8	13	19
	400	0	0	0	1	1	1	3	4	6	7	12	17
	500	0	0	0	1	1	1	2	3	5	6	10	14
	600	0	0	0	1	1	1	1	3	4	5	8	12
	700	0	0	0	0	1	1	1	2	3	4	7	10
	750	0	0	0	0	1	1	1	2	3	4	7	10
	800	0	0	0	0	1	1	1	2	3	4	6	9
	900	0	0	0	0	1	1	1	1	3	4	6	8

	1000	0	0	0	0	0	1	1	1	2	3	5	8
	1250	0	0	0	0	0	1	1	1	1	2	4	6
	1500	0	0	0	0	0	1	1	1	1	2	3	5
	1750	0	0	0	0	0	0	1	1	1	1	3	4
	2000	0	0	0	0	0	0	1	1	1	1	3	4
THHN, THWN, THWN-2	14	13	22	36	63	85	140	200	309	412	531	833	1202
	12	9	16	26	46	62	102	146	225	301	387	608	877
	10	6	10	17	29	39	64	92	142	189	244	383	552
	8	3	6	9	16	22	37	53	82	109	140	221	318
	6	2	4	7	12	16	27	38	59	79	101	159	230
	4	1	2	4	7	10	16	23	36	48	62	98	141
	3	1	1	3	6	8	14	20	31	41	53	83	120
	2	1	1	3	5	7	11	17	26	34	44	70	100
	1	1	1	1	4	5	8	12	19	25	33	51	74
	1/0	1	1	1	3	4	7	10	16	21	27	43	63
	2/0	0	1	1	2	3	6	8	13	18	23	36	52
	3/0	0	1	1	1	3	5	7	11	15	19	30	43
	4/0	0	1	1	1	2	4	6	9	12	16	25	36
	250	0	0	1	1	1	3	5	7	10	13	20	29
	300	0	0	1	1	1	3	4	6	8	11	17	25
	350	0	0	1	1	1	2	3	5	7	10	15	22
	400	0	0	1	1	1	2	3	5	7	8	13	20
	500	0	0	0	1	1	1	2	4	5	7	11	16
	600	0	0	0	1	1	1	1	3	4	6	9	13
	700	0	0	0	1	1	1	1	3	4	5	8	11
	750	0	0	0	0	1	1	1	3	4	5	7	11
	800	0	0	0	0	1	1	1	2	3	4	7	10
	900	0	0	0	0	1	1	1	2	3	4	6	9
	1000	0	0	0	0	1	1	1	1	3	4	6	8

Table C8 Continued

		Conductors											
		Metric Designator (Trade Size)											
Type	*Conductor Size (AWG/kcmil)*	*16 (½)*	*21 (¾)*	*27 (1)*	*35 (1¼)*	*41 (1½)*	*53 (2)*	*63 (2½)*	*78 (3)*	*91 (3½)*	*103 (4)*	*129 (5)*	*155 (6)*
FEP, FEPB, PFA, PFAH, TFE	14	12	22	35	61	83	136	194	300	400	515	808	1166
	12	9	16	26	44	60	99	142	219	292	376	590	851
	10	6	11	18	32	43	71	102	157	209	269	423	610
	8	3	6	10	18	25	41	58	90	120	154	242	350
	6	2	4	7	13	17	29	41	64	85	110	172	249
	4	1	3	5	9	12	20	29	44	59	77	120	174
	3	1	2	4	7	10	17	24	37	50	64	100	145
	2	1	1	3	6	8	14	20	31	41	53	83	120
PFA, PFAH, TFE	1	1	1	2	4	6	9	14	21	28	37	57	83
PFA, PFAH, TFE, Z	1/0	1	1	1	3	5	8	11	18	24	30	48	69
	2/0	1	1	1	3	4	6	9	14	19	25	40	57
	3/0	0	1	1	2	3	5	8	12	16	21	33	47
	4/0	0	1	1	1	2	4	6	10	13	17	27	39

Z	14	15	26	42	73	100	164	234	361	482	621	974	1405
	12	10	18	30	52	71	116	166	256	342	440	691	997
	10	6	11	18	32	43	71	102	157	209	269	423	610
	8	4	7	11	20	27	45	64	99	132	170	267	386
	6	3	5	8	14	19	31	45	69	93	120	188	271
	4	1	3	5	9	13	22	31	48	64	82	129	186
	3	1	2	4	7	9	16	22	35	47	60	94	136
	2	1	1	3	6	8	13	19	29	39	50	78	113
	1	1	1	2	5	6	10	15	23	31	40	63	92
XHH, XHHW, XHHW-2, ZW	14	9	15	25	44	59	98	140	216	288	370	581	839
	12	7	12	19	33	45	75	107	165	221	284	446	644
	10	5	9	14	25	34	56	80	123	164	212	332	480
	8	3	5	8	14	19	31	44	68	91	118	185	267
	6	1	3	6	10	14	23	33	51	68	87	137	197
	4	1	2	4	7	10	16	24	37	49	63	99	143
	3	1	1	3	6	8	14	20	31	41	53	84	121
	2	1	1	3	5	7	12	17	26	35	45	70	101
XHH, XHHW, XHHW-2	1	1	1	1	4	5	9	12	19	26	33	52	76
	1/0	1	1	1	3	4	7	10	16	22	28	44	64
	2/0	0	1	1	2	3	6	9	13	18	23	37	53
	3/0	0	1	1	1	3	5	7	11	15	19	30	44
	4/0	0	1	1	1	2	4	6	9	12	16	25	36
	250	0	0	1	1	1	3	5	7	10	13	20	30
	300	0	0	1	1	1	3	4	6	9	11	18	25
	350	0	0	1	1	1	2	3	6	7	10	15	22
	400	0	0	1	1	1	2	3	5	7	9	14	20
	500	0	0	0	1	1	1	2	4	5	7	11	16

Table C8 *Continued*

Conductors

Type	*Conductor Size (AWG/kcmil)*	*Metric Designator (Trade Size)* 16 (½)	21 (¾)	27 (1)	35 (1¼)	41 (1½)	53 (2)	63 (2½)	78 (3)	91 (3½)	103 (4)	129 (5)	155 (6)
XHH, XHHW, XHHW-2	600	0	0	0	1	1	1	1	3	4	6	9	13
	700	0	0	0	1	1	1	1	3	4	5	8	11
	750	0	0	0	0	1	1	1	3	4	5	7	11
	800	0	0	0	0	1	1	1	2	3	4	7	10
	900	0	0	0	0	1	1	1	2	3	4	6	9
	1000	0	0	0	0	1	1	1	1	3	4	6	8
	1250	0	0	0	0	0	1	1	1	2	3	4	6
	1500	0	0	0	0	0	1	1	1	1	2	4	5
	1750	0	0	0	0	0	0	1	1	1	1	3	5
	2000	0	0	0	0	0	0	1	1	1	1	3	4

Fixture Wires

Type	Conductor Size (AWG/kcmil)	Metric Designator (Trade Size) 16 (½)	21 (¾)	27 (1)	35 (1¼)	41 (1½)	53 (2)
FFH-2, RFH-2, RFHH-3	18	8	15	24	42	57	94
	16	7	12	20	35	48	79
SF-2, SFF-2	18	11	19	31	53	72	118
	16	9	15	25	44	59	98
	14	7	12	20	35	48	79
SF-1, SFF-1	18	19	33	54	94	127	209
RFH-1, RFHH-2, TF, TFF, XF, XFF	18	14	25	40	69	94	155
RFHH-2, TF, TFF, XF, XFF	16	11	20	32	56	76	125
XF, XFF	14	9	15	25	44	59	98
TFN, TFFN	18	23	40	64	111	150	248
	16	17	30	49	84	115	189
PF, PFF, PGF, PGFF, PAF, PTF, PTFF, PAFF	18	21	38	61	105	143	235
	16	16	29	47	81	110	181
	14	12	22	35	61	83	136
HF, HFF, ZF, ZFF, ZHF	18	28	48	79	135	184	303
	16	20	36	58	100	136	223
	14	15	26	42	73	100	164
KF-2, KFF-2	18	40	71	114	197	267	439
	16	28	50	80	138	188	310
	14	19	34	55	95	129	213
	12	13	23	38	65	89	146
	10	9	15	25	44	59	98
KF-1, KFF-1	18	48	84	136	235	318	524
	16	34	59	96	165	224	368
	14	23	40	64	111	150	248
	12	15	26	42	73	100	164
	10	10	17	28	48	65	107
XF, XFF	12	5	8	13	23	32	52
	10	3	6	10	18	25	41

Note: This table is for concentric stranded conductors only. For compact stranded conductors, Table C8(A) should be used.

*Types RHH, RHW, and RHW-2 without outer covering.

Table C8(A) Maximum Number of Compact Conductors in Rigid Metal Conduit (RMC) (*Based on Table 1, Chapter 9*)

Compact Conductors													
		Metric Designator (Trade Size)											
Type	*Conductor Size (AWG/kcmil)*	*16 (½)*	*21 (¾)*	*27 (1)*	*35 (1¼)*	*41 (1½)*	*53 (2)*	*63 (2½)*	*78 (3)*	*91 (3½)*	*103 (4)*	*129 (5)*	*155 (6)*
THW, THW-2, THHW	8	2	4	7	12	16	26	38	59	78	101	158	228
	6	1	3	5	9	12	20	29	45	60	78	122	176
	4	1	2	4	7	9	15	22	34	45	58	91	132
	2	1	1	3	5	7	11	16	25	33	43	67	97
	1	1	1	1	3	5	8	11	17	23	30	47	68
	1/0	1	1	1	3	4	7	10	15	20	26	41	59
	2/0	0	1	1	2	3	6	8	13	17	22	34	50
	3/0	0	1	1	1	3	5	7	11	14	19	29	42
	4/0	0	1	1	1	2	4	6	9	12	15	24	35
	250	0	0	1	1	1	3	4	7	9	12	19	28
	300	0	0	1	1	1	3	4	6	8	11	17	24
	350	0	0	1	1	1	2	3	5	7	9	15	22
	400	0	0	1	1	1	1	3	5	7	8	13	20
	500	0	0	0	1	1	1	3	4	5	7	11	17
	600	0	0	0	1	1	1	1	3	4	6	9	13
	700	0	0	0	1	1	1	1	3	4	5	8	12
	750	0	0	0	0	1	1	1	3	4	5	7	11
	1000	0	0	0	0	1	1	1	1	3	4	6	9

THHN, THWN, THWN-2	8	—	—	—	—	—	—	—	—	—	—	—	—
	6	2	5	8	13	18	30	43	66	88	114	179	258
	4	1	3	5	8	11	18	26	41	55	70	110	159
	2	1	1	3	6	8	13	19	29	39	50	79	114
	1	1	1	2	4	6	10	14	22	29	38	60	86
	1/0	1	1	1	4	5	8	12	19	25	32	51	73
	2/0	1	1	1	3	4	7	10	15	21	26	42	60
	3/0	0	1	1	2	3	6	8	13	17	22	35	51
	4/0	0	1	1	1	3	5	7	10	14	18	29	42
	250	0	1	1	1	2	4	5	8	11	14	23	33
	300	0	0	1	1	1	3	4	7	10	12	20	28
	350	0	0	1	1	1	3	4	6	8	11	17	25
	400	0	0	1	1	1	2	3	5	7	10	15	22
	500	0	0	0	1	1	1	3	5	6	8	13	19
	600	0	0	0	1	1	1	2	4	5	6	10	15
	700	0	0	0	1	1	1	1	3	4	6	9	13
	750	0	0	0	1	1	1	1	3	4	5	9	13
	1000	0	0	0	0	1	1	1	2	3	4	6	9

Table C8(A) *Continued*

		Compact Conductors											
		Metric Designator (Trade Size)											
Type	*Conductor Size (AWG/kcmil)*	*16 (½)*	*21 (¾)*	*27 (1)*	*35 (1¼)*	*41 (1½)*	*53 (2)*	*63 (2½)*	*78 (3)*	*91 (3½)*	*103 (4)*	*129 (5)*	*155 (6)*
XHHW, XHH2-2	8	3	5	9	15	21	34	49	76	101	130	205	296
	6	2	4	6	11	15	25	36	56	75	97	152	220
	4	1	3	5	8	11	18	26	41	55	70	110	159
	2	1	1	3	6	8	13	19	29	39	50	79	114
	1	1	1	2	4	6	10	14	22	29	38	60	86
	1/0	1	1	1	4	5	8	12	19	25	32	51	73
	2/0	1	1	1	3	4	7	10	16	21	27	43	62
	3/0	0	1	1	2	3	6	8	13	17	22	35	51
	4/0	0	1	1	1	3	5	7	11	14	19	29	42
	250	0	1	1	1	2	4	5	8	11	15	23	34
	300	0	0	1	1	1	3	5	7	10	13	20	29
	350	0	0	1	1	1	3	4	6	9	11	18	25
	400	0	0	1	1	1	2	4	6	8	10	16	23
	500	0	0	0	1	1	1	3	5	6	8	13	19
	600	0	0	0	1	1	1	2	4	5	7	10	15
	700	0	0	0	1	1	1	1	3	4	6	9	13
	750	0	0	0	1	1	1	1	3	4	5	8	12
	1000	0	0	0	0	1	1	1	2	3	4	7	10

Definition: *Compact stranding* is the result of a manufacturing process where the standard conductor is compressed to the extent that the interstices (voids between strand wires) are virtually eliminated.

Table C9 Maximum Number of Conductors or Fixture Wires in Rigid PVC Conduit, Schedule 80 (*Based on Table 1, Chapter 9*)

		Conductors											
	Conductor Size	*Metric Designator (Trade Size)*											
Type	*(AWG/kcmil)*	*16 (½)*	*21 (¾)*	*27 (1)*	*35 (1¼)*	*41 (1½)*	*53 (2)*	*63 (2½)*	*78 (3)*	*91 (3½)*	*103 (4)*	*129 (5)*	*155 (6)*
RHH, RHW, RHW-2	14	3	5	9	17	23	39	56	88	118	153	243	349
	12	2	4	7	14	19	32	46	73	98	127	202	290
	10	1	3	6	11	15	26	37	59	79	103	163	234
	8	1	1	3	6	8	13	19	31	41	54	85	122
	6	1	1	2	4	6	11	16	24	33	43	68	98
	4	1	1	1	3	5	8	12	19	26	33	53	77
	3	0	1	1	3	4	7	11	17	23	29	47	67
	2	0	1	1	3	4	6	9	14	20	25	41	58
	1	0	1	1	1	2	4	6	9	13	17	27	38
	1/0	0	0	1	1	1	3	5	8	11	15	23	33
	2/0	0	0	1	1	1	3	4	7	10	13	20	29
	3/0	0	0	1	1	1	3	4	6	8	11	17	25
	4/0	0	0	0	1	1	2	3	5	7	9	15	21
	250	0	0	0	1	1	1	2	4	5	7	11	16
	300	0	0	0	1	1	1	2	3	5	6	10	14
	350	0	0	0	1	1	1	1	3	4	5	9	13
	400	0	0	0	0	1	1	1	3	4	5	8	12
	500	0	0	0	0	1	1	1	2	3	4	7	10

Table C9 Continued

		Conductors											
		Metric Designator (Trade Size)											
Type	Conductor Size (AWG/kcmil)	16 (½)	21 (¾)	27 (1)	35 (1¼)	41 (1½)	53 (2)	63 (2½)	78 (3)	91 (3½)	103 (4)	129 (5)	155 (6)
RHH, RHW, RHW-2	600	0	0	0	0	0	1	1	1	3	3	6	8
	700	0	0	0	0	0	1	1	1	2	3	5	7
	750	0	0	0	0	0	1	1	1	2	3	5	7
	800	0	0	0	0	0	1	1	1	2	3	4	7
	1000	0	0	0	0	0	1	1	1	1	2	4	5
	1250	0	0	0	0	0	0	1	1	1	1	3	4
	1500	0	0	0	0	0	0	1	1	1	1	2	4
	1750	0	0	0	0	0	0	0	1	1	1	2	3
	2000	0	0	0	0	0	0	0	1	1	1	1	3
TW, THHW, THW, THW-2	14	6	11	20	35	49	82	118	185	250	324	514	736
	12	5	9	15	27	38	63	91	142	192	248	394	565
	10	3	6	11	20	28	47	67	106	143	185	294	421
	8	1	3	6	11	15	26	37	59	79	103	163	234
RHH*, RHW*, RHW-2*	14	4	8	13	23	32	55	79	123	166	215	341	490
	12	3	6	10	19	26	44	63	99	133	173	274	394
	10	2	5	8	15	20	34	49	77	104	135	214	307
	8	1	3	5	9	12	20	29	46	62	81	128	184

RHH*,	6	1	1	3	7	9	16	22	35	48	62	98	141
RHW*,	4	1	1	3	5	7	12	17	26	35	46	73	105
RHW-2*,	3	1	1	2	4	6	10	14	22	30	39	63	90
TW,	2	1	1	1	3	5	8	12	19	26	33	53	77
THW,	1	0	1	1	2	3	6	8	13	18	23	37	54
THHW,	1/0	0	1	1	1	3	5	7	11	15	20	32	46
THW-2	2/0	0	1	1	1	2	4	6	10	13	17	27	39
	3/0	0	0	1	1	1	3	5	8	11	14	23	33
	4/0	0	0	1	1	1	3	4	7	9	12	19	27
	250	0	0	0	1	1	2	3	5	7	9	15	22
	300	0	0	0	1	1	1	3	5	6	8	13	19
	350	0	0	0	1	1	1	2	4	6	7	12	17
	400	0	0	0	1	1	1	2	4	5	7	10	15
	500	0	0	0	1	1	1	1	3	4	5	9	13
	600	0	0	0	0	1	1	1	2	3	4	7	10
	700	0	0	0	0	1	1	1	2	3	4	6	9
	750	0	0	0	0	0	1	1	1	3	4	6	8
	800	0	0	0	0	0	1	1	1	3	3	6	8
	900	0	0	0	0	0	1	1	1	2	3	5	7
	1000	0	0	0	0	0	1	1	1	2	3	5	7
	1250	0	0	0	0	0	1	1	1	1	2	4	5
	1500	0	0	0	0	0	0	1	1	1	1	3	4
	1750	0	0	0	0	0	0	1	1	1	1	3	4
	2000	0	0	0	0	0	0	0	1	1	1	2	3

Table C9 *Continued*

		Conductors											
		Metric Designator (Trade Size)											
Type	*Conductor Size (AWG/kcmil)*	*16 (½)*	*21 (¾)*	*27 (1)*	*35 (1¼)*	*41 (1½)*	*53 (2)*	*63 (2½)*	*78 (3)*	*91 (3½)*	*103 (4)*	*129 (5)*	*155 (6)*
THHN, THWN, THWN-2	14	9	17	28	51	70	118	170	265	358	464	736	1055
	12	6	12	20	37	51	86	124	193	261	338	537	770
	10	4	7	13	23	32	54	78	122	164	213	338	485
	8	2	4	7	13	18	31	45	70	95	123	195	279
	6	1	3	5	9	13	22	32	51	68	89	141	202
	4	1	1	3	6	8	14	20	31	42	54	86	124
	3	1	1	3	5	7	12	17	26	35	46	73	105
	2	1	1	2	4	6	10	14	22	30	39	61	88
	1	0	1	1	3	4	7	10	16	22	29	45	65
	1/0	0	1	1	2	3	6	9	14	18	24	38	55
	2/0	0	1	1	1	3	5	7	11	15	20	32	46
	3/0	0	1	1	1	2	4	6	9	13	17	26	38
	4/0	0	0	1	1	1	3	5	8	10	14	22	31
	250	0	0	1	1	1	3	4	6	8	11	18	25
	300	0	0	0	1	1	2	3	5	7	9	15	22
	350	0	0	0	1	1	1	3	5	6	8	13	19
	400	0	0	0	1	1	1	3	4	6	7	12	17
	500	0	0	0	1	1	1	2	3	5	6	10	14

	600	0	0	0	0	1	1	1	3	4	5	8	12
	700	0	0	0	0	1	1	1	2	3	4	7	10
	750	0	0	0	0	1	1	1	2	3	4	7	9
	800	0	0	0	0	1	1	1	2	3	4	6	9
	900	0	0	0	0	0	1	1	1	3	3	6	8
	1000	0	0	0	0	0	1	1	1	2	3	5	7
FEP,	14	8	16	27	49	68	115	164	257	347	450	714	1024
FEPB,	12	6	12	20	36	50	84	120	188	253	328	521	747
PFA,	10	4	8	14	26	36	60	86	135	182	235	374	536
PFAH,	8	2	5	8	15	20	34	49	77	104	135	214	307
TFE	6	1	3	6	10	14	24	35	55	74	96	152	218
	4	1	2	4	7	10	17	24	38	52	67	106	153
	3	1	1	3	6	8	14	20	32	43	56	89	127
	2	1	1	3	5	7	12	17	26	35	46	73	105
PFA, PFAH, TFE	1	1	1	1	3	5	8	11	18	25	32	51	73
PFA,	1/0	0	1	1	3	4	7	10	15	20	27	42	61
PFAH,	2/0	0	1	1	2	3	5	8	12	17	22	35	50
TFE, Z	3/0	0	1	1	1	2	4	6	10	14	18	29	41
	4/0	0	0	1	1	1	4	5	8	11	15	24	34
Z	14	10	19	33	59	82	138	198	310	418	542	860	1233
	12	7	14	23	42	58	98	141	220	297	385	610	875
	10	4	8	14	26	36	60	86	135	182	235	374	536
	8	3	5	9	16	22	38	54	85	115	149	236	339
	6	2	4	6	11	16	26	38	60	81	104	166	238
	4	1	2	4	8	11	18	26	41	55	72	114	164
	3	1	2	3	5	8	13	19	30	40	52	83	119
	2	1	1	2	5	6	11	16	25	33	43	69	99
	1	0	1	2	4	5	9	13	20	27	35	56	80

Table C9 *Continued*

		Conductors											
		Metric Designator (Trade Size)											
Type	*Conductor Size (AWG/kcmil)*	*16 (½)*	*21 (¾)*	*27 (1)*	*35 (1¼)*	*41 (1½)*	*53 (2)*	*63 (2½)*	*78 (3)*	*91 (3½)*	*103 (4)*	*129 (5)*	*155 (6)*
XHH,	14	6	11	20	35	49	82	118	185	250	324	514	736
XHHW,	12	5	9	15	27	38	63	91	142	192	248	394	565
XHHW-2,	10	3	6	11	20	28	47	67	106	143	185	294	421
ZW	8	1	3	6	11	15	26	37	59	79	103	163	234
	6	1	2	4	8	11	19	28	43	59	76	121	173
	4	1	1	3	6	8	14	20	31	42	55	87	125
	3	1	1	3	5	7	12	17	26	36	47	74	106
	2	1	1	2	4	6	10	14	22	30	39	62	89

XHH,	1	0	1	1	3	4	7	10	16	22	29	46	66
XHHW,	1/0	0	1	1	2	3	6	9	14	19	24	39	56
XHHW-2	2/0	0	1	1	1	3	5	7	11	16	20	32	46
	3/0	0	1	1	1	2	4	6	9	13	17	27	38
	4/0	0	0	1	1	1	3	5	8	11	14	22	32
	250	0	0	1	1	1	3	4	6	9	11	18	26
	300	0	0	1	1	1	2	3	5	7	10	15	22
	350	0	0	0	1	1	1	3	5	6	8	14	20
	400	0	0	0	1	1	1	3	4	6	7	12	17
	500	0	0	0	1	1	1	2	3	5	6	10	14
	600	0	0	0	0	1	1	1	3	4	5	8	11
	700	0	0	0	0	1	1	1	2	3	4	7	10
	750	0	0	0	0	1	1	1	2	3	4	6	9
	800	0	0	0	0	1	1	1	1	3	4	6	9
	900	0	0	0	0	0	1	1	—	3	3	5	8
	1000	0	0	0	0	0	1	1	1	2	3	5	7
	1250	0	0	0	0	0	1	1	1	1	2	4	6
	1500	0	0	0	0	0	0	1	1	1	1	3	5
	1750	0	0	0	0	0	0	1	1	1	1	3	4
	2000	0	0	0	0	0	0	1	1	1	1	2	4

Fixture Wires

Type	Conductor Size (AWG/kcmil)	Metric Designator (Trade Size) 16 (½)	21 (¾)	27 (1)	35 (1¼)	41 (1½)	53 (2)
FFH-2, RFH-2, RFHH-3	18	6	11	19	34	47	79
	16	5	9	16	28	39	67
SF-2, SFF-2	18	7	14	24	43	59	100
	16	6	11	20	35	49	82
	14	5	9	16	28	39	67
SF-1, SFF-1	18	13	25	42	76	105	177
RFH-1, RFHH-2, TF, TFF, XF, XFF	18	10	18	31	56	77	130
RFHH-2, TF, TFF, XF, XFF	16	8	15	25	45	62	105
XF, XFF	14	6	11	20	35	49	82
TFN, TFFN	18	16	29	50	90	124	209
	16	12	22	38	68	95	159
PF, PFF, PGF, PGFF, PAF, PTF, PTFF, PAFF	18	15	28	47	85	118	198
	16	11	22	36	66	91	153
	14	8	16	27	49	68	115
HF, HFF, ZF, ZFF, ZHF	18	19	36	61	110	152	255
	16	14	27	45	81	112	188
	14	10	19	33	59	82	138
KF-2, KFF-2	18	28	53	88	159	220	371
	16	19	37	62	112	155	261
	14	13	25	43	77	107	179
	12	9	17	29	53	73	123
	10	6	11	20	35	49	82
KF-1, KFF-1	18	33	63	106	190	263	442
	16	23	44	74	133	185	310
	14	16	29	50	90	124	209
	12	10	19	33	59	82	138
	10	7	13	21	39	54	90
XF, XFF	12	3	6	10	19	26	44
	10	2	5	8	15	20	34

Note: This table is for concentric stranded conductors only. For compact stranded conductors, Table C9(A) should be used.

*Types RHH, RHW, and RHW-2 without outer covering.

Table C9(A) Maximum Number of Compact Conductors in Rigid PVC Conduit, Schedule 80 (*Based on Table 1, Chapter 9*)

		Compact Conductors											
	Conductor	*Metric Designator (Trade Size)*											
Type	*Size (AWG/kcmil)*	*16 (½)*	*21 (¾)*	*27 (1)*	*35 (1¼)*	*41 (1½)*	*53 (2)*	*63 (2½)*	*78 (3)*	*91 (3½)*	*103 (4)*	*129 (5)*	*155 (6)*
THW, THW-2, THHW	8	1	3	5	9	13	22	32	50	68	88	140	200
	6	1	2	4	7	10	17	25	39	52	68	108	155
	4	1	1.	3	5	7	13	18	29	39	51	81	116
	2	1	1	1	4	5	9	13	21	29	37	60	85
	1	0	1	1	3	4	6	9	15	20	26	42	60
	1/0	0	1	1	2	3	6	8	13	17	23	36	52
	2/0	0	1	1	1	3	5	7	11	15	19	30	44
	3/0	0	0	1	1	2	4	6	9	12	16	26	37
	4/0	0	0	1	1	1	3	5	8	10	13	22	31
	250	0	0	1	1	1	2	4	6	8	11	17	25
	300	0	0	0	1	1	2	3	5	7	9	15	21
	350	0	0	0	1	1	1	3	5	6	8	13	19
	400	0	0	0	1	1	1	3	4	6	7	12	17
	500	0	0	0	1	1	1	2	3	5	6	10	14
	600	0	0	0	0	1	1	1	3	4	5	8	12
	700	0	0	0	0	1	1	1	2	3	4	7	10
	750	0	0	0	0	1	1	1	2	3	4	7	10
	1000	0	0	0	0	0	1	1	1	2	3	5	8

Table C9(A) *Continued*

		Compact Conductors											
	Conductor	*Metric Designator (Trade Size)*											
Type	*Size (AWG/kcmil)*	*16 (½)*	*21 (¾)*	*27 (1)*	*35 (1¼)*	*41 (1½)*	*53 (2)*	*63 (2½)*	*78 (3)*	*91 (3½)*	*103 (4)*	*129 (5)*	*155 (6)*
THHN, THWN, THWN-2	8	—	—	—	—	—	—	—	—	—	—	—	—
	6	1	3	6	11	15	25	36	57	77	99	158	226
	4	1	1	3	6	9	15	22	35	47	61	98	140
	2	1	1	2	5	6	11	16	25	34	44	70	100
	1	1	1	1	3	5	8	12	19	25	33	53	75
	1/0	0	1	1	3	4	7	10	16	22	28	45	64
	2/0	0	1	1	2	3	6	8	13	18	23	37	53
	3/0	0	1	1	1	3	5	7	11	15	19	31	44
	4/0	0	0	1	1	2	4	6	9	12	16	25	37
	250	0	0	1	1	1	3	4	7	10	12	20	29
	300	0	0	1	1	1	3	4	6	8	11	17	25
	350	0	0	0	1	1	2	3	5	7	9	15	22
	400	0	0	0	1	1	1	3	5	6	8	13	19
	500	0	0	0	1	1	1	2	4	5	7	11	16
	600	0	0	0	1	1	1	1	3	4	6	9	13
	700	0	0	0	0	1	1	1	3	4	5	8	12
	750	0	0	0	0	1	1	1	3	4	5	8	11
	1000	0	0	0	0	0	1	1	1	3	3	5	8

XHHW, XHHW-2	8	1	4	7	12	17	29	42	65	88	114	181	260
	6	1	3	5	9	13	21	31	48	65	85	134	193
	4	1	1	3	6	9	15	22	35	47	61	98	140
	2	1	1	2	5	6	11	16	25	34	44	70	100
	1	1	1	1	3	5	8	12	19	25	33	53	75
	1/0	0	1	1	3	4	7	10	16	22	28	45	64
	2/0	0	1	1	2	3	6	8	13	18	24	38	54
	3/0	0	1	1	1	3	5	7	11	15	19	31	44
	4/0	0	0	1	1	2	4	6	9	12	16	26	37
	250	0	0	1	1	1	3	5	7	10	13	21	30
	300	0	0	1	1	1	3	4	6	8	11	17	25
	350	0	0	1	1	1	2	3	5	7	10	15	22
	400	0	0	0	1	1	1	3	5	7	9	14	20
	500	0	0	0	1	1	1	2	4	5	7	11	17
	600	0	0	0	1	1	1	1	3	4	6	9	13
	700	0	0	0	0	1	1	1	3	4	5	8	12
	750	0	0	0	0	1	1	1	2	3	5	7	11
	1000	0	0	0	0	0	1	1	1	3	3	6	8

Definition: *Compact stranding* is the result of a manufacturing process where the standard conductor is compressed to the extent that the interstices (voids between strand wires) are virtually eliminated.

Table C10 Maximum Number of Conductors or Fixture Wires in Rigid PVC Conduit, Schedule 40 and HDPE Conduit (*Based on Table 1, Chapter 9*)

		Conductors											
		Metric Designator (Trade Size)											
Type	*Conductor Size (AWG/kcmil)*	*16 (½)*	*21 (¾)*	*27 (1)*	*35 (1¼)*	*41 (1½)*	*53 (2)*	*63 (2½)*	*78 (3)*	*91 (3½)*	*103 (4)*	*129 (5)*	*155 (6)*
RHH, RHW, RHW-2	14	4	7	11	20	27	45	64	99	133	171	269	390
	12	3	5	9	16	22	37	53	82	110	142	224	323
	10	2	4	7	13	18	30	43	66	89	115	181	261
	8	1	2	4	7	9	15	22	35	46	60	94	137
	6	1	1	3	5	7	12	18	28	37	48	76	109
	4	1	1	2	4	6	10	14	22	29	37	59	85
	3	1	1	1	4	5	8	12	19	25	33	52	75
	2	1	1	1	3	4	7	10	16	22	28	45	65
	1	0	1	1	1	3	5	7	11	14	19	29	43
	1/0	0	1	1	1	2	4	6	9	13	16	26	37
	2/0	0	0	1	1	1	3	5	8	11	14	22	32
	3/0	0	0	1	1	1	3	4	7	9	12	19	28
	4/0	0	0	1	1	1	2	4	6	8	10	16	24
	250	0	0	0	1	1	1	3	4	6	8	12	18
	300	0	0	0	1	1	1	2	4	5	7	11	16
	350	0	0	0	1	1	1	2	3	5	6	10	14
	400	0	0	0	1	1	1	1	3	4	6	9	13
	500	0	0	0	0	1	1	1	3	4	5	8	11

	600	0	0	0	0	1	1	1	2	3	4	6	9
	700	0	0	0	0	0	1	1	1	3	3	6	8
	750	0	0	0	0	0	1	1	1	2	3	5	8
	800	0	0	0	0	0	1	1	1	2	3	5	7
	900	0	0	0	0	0	1	1	1	2	3	5	7
	1000	0	0	0	0	0	1	1	1	1	3	4	6
	1250	0	0	0	0	0	0	1	1	1	1	3	5
	1500	0	0	0	0	0	0	1	1	1	1	3	4
	1750	0	0	0	0	0	0	1	1	1	1	2	3
	2000	0	0	0	0	0	0	0	1	1	1	2	3
TW,	14	8	14	24	42	57	94	135	209	280	361	568	822
THHW,	12	6	11	18	32	44	72	103	160	215	277	436	631
THW,	10	4	8	13	24	32	54	77	119	160	206	325	470
THW-2	8	2	4	7	13	18	30	43	66	89	115	181	261
RHH*,	14	5	9	16	28	38	63	90	139	186	240	378	546
RHW*,	12	4	8	12	22	30	50	72	112	150	193	304	439
RHW-2*	10	3	6	10	17	24	39	56	87	117	150	237	343
	8	1	3	6	10	14	23	33	52	70	90	142	205
TW, THW,	6	1	2	4	8	11	18	26	40	53	69	109	157
THHW,	4	1	1	3	6	8	13	19	30	40	51	81	117
THW-2	3	1	1	3	5	7	11	16	25	34	44	69	100
	2	1	1	2	4	6	10	14	22	29	37	59	85
	1	0	1	1	3	4	7	10	15	20	26	41	60
	1/0	0	1	1	2	3	6	8	13	17	22	35	51
	2/0	0	1	1	1	3	5	7	11	15	19	30	43
	3/0	0	1	1	1	2	4	6	9	12	16	25	36
	4/0	0	0	1	1	1	3	5	8	10	13	21	30

Table C10 *Continued*

		Conductors											
		Metric Designator (Trade Size)											
Type	*Conductor Size (AWG/kcmil)*	*16 (½)*	*21 (¾)*	*27 (1)*	*35 (1¼)*	*41 (1½)*	*53 (2)*	*63 (2½)*	*78 (3)*	*91 (3½)*	*103 (4)*	*129 (5)*	*155 (6)*
TW, THW, THHW, THW-2	250	0	0	1	1	1	3	4	6	8	11	17	25
	300	0	0	1	1	1	2	3	5	7	9	15	21
	350	0	0	0	1	1	1	3	5	6	8	13	19
	400	0	0	0	1	1	1	3	4	6	7	12	17
	500	0	0	0	1	1	1	2	3	5	6	10	14
	600	0	0	0	0	1	1	1	3	4	5	8	11
	700	0	0	0	0	1	1	1	2	3	4	7	10
	750	0	0	0	0	1	1	1	2	3	4	6	10
	800	0	0	0	0	1	1	1	2	3	4	6	9
	900	0	0	0	0	0	1	1	1	3	3	6	8
	1000	0	0	0	0	0	1	1	1	2	3	5	7
	1250	0	0	0	0	0	1	1	1	1	2	4	6
	1500	0	0	0	0	0	1	1	1	1	1	3	5
	1750	0	0	0	0	0	0	1	1	1	1	3	4
	2000	0	0	0	0	0	0	1	1	1	1	3	4
THHN, THWN, THWN-2	14	11	21	34	60	82	135	193	299	401	517	815	1178
	12	8	15	25	43	59	99	141	218	293	377	594	859
	10	5	9	15	27	37	62	89	137	184	238	374	541
	8	3	5	9	16	21	36	51	79	106	137	216	312
	6	1	4	6	11	15	26	37	57	77	99	156	225

	4	1	2	4	7	9	16	22	35	47	61	96	138
	3	1	1	3	6	8	13	19	30	40	51	81	117
	2	1	1	3	5	7	11	16	25	33	43	68	98
	1	1	1	1	3	5	8	12	18	25	32	50	73
	1/0	1	1	1	3	4	7	10	15	21	27	42	61
	2/0	0	1	1	2	3	6	8	13	17	22	35	51
	3/0	0	1	1	1	3	5	7	11	14	18	29	42
	4/0	0	1	1	1	2	4	6	9	12	15	24	35
	250	0	0	1	1	1	3	4	7	10	12	20	28
	300	0	0	1	1	1	3	4	6	8	11	17	24
	350	0	0	1	1	1	2	3	5	7	9	15	21
	400	0	0	0	1	1	1	3	5	6	8	13	19
	500	0	0	0	1	1	1	2	4	5	7	11	16
	600	0	0	0	1	1	1	1	3	4	5	9	13
	700	0	0	0	0	1	1	1	3	4	5	8	11
	750	0	0	0	0	1	1	1	2	3	4	7	11
	800	0	0	0	0	1	1	1	2	3	4	7	10
	900	0	0	0	0	1	1	1	2	3	4	6	9
	1000	0	0	0	0	0	1	1	1	3	3	6	8
FEP,	14	11	20	33	58	79	131	188	290	389	502	790	1142
FEPB,	12	8	15	24	42	58	96	137	212	284	366	577	834
PFA,	10	6	10	17	30	41	69	98	152	204	263	414	598
PFAH,	8	3	6	10	17	24	39	56	87	117	150	237	343
TFE	6	2	4	7	12	17	28	40	62	83	107	169	244
	4	1	3	5	8	12	19	28	43	58	75	118	170
	3	1	2	4	7	10	16	23	36	48	62	98	142
	2	1	1	3	6	8	13	19	30	40	51	81	117

Table C10 *Continued*

		Conductors											
		Metric Designator (Trade Size)											
Type	*Conductor Size (AWG/kcmil)*	*16 (½)*	*21 (¾)*	*27 (1)*	*35 (1¼)*	*41 (1½)*	*53 (2)*	*63 (2½)*	*78 (3)*	*91 (3½)*	*103 (4)*	*129 (5)*	*155 (6)*
PFA, PFAH, TFE	1	1	1	2	4	5	9	13	20	28	36	56	81
PFA, PFAH, TFE, Z	1/0	1	1	1	3	4	8	11	17	23	30	47	68
	2/0	0	1	1	3	4	6	9	14	19	24	39	56
	3/0	0	1	1	2	3	5	7	12	16	20	32	46
	4/0	0	1	1	1	2	4	6	9	13	16	26	38
Z	14	13	24	40	70	95	158	226	350	469	605	952	1376
	12	9	17	28	49	68	112	160	248	333	429	675	976
	10	6	10	17	30	41	69	98	152	204	263	414	598
	8	3	6	11	19	26	43	62	96	129	166	261	378
	6	2	4	7	13	18	30	43	67	90	116	184	265
	4	1	3	5	9	12	21	30	46	62	80	126	183
	3	1	2	4	6	9	15	22	34	45	58	92	133
	2	1	1	3	5	7	12	18	28	38	49	77	111
	1	1	1	2	4	6	10	14	23	30	39	62	90

XHH, XHHW, XHHW-2, ZW	14	8	14	24	42	57	94	135	209	280	361	568	822
	12	6	11	18	32	44	72	103	160	215	277	436	631
	10	4	8	13	24	32	54	77	119	160	206	325	470
	8	2	4	7	13	18	30	43	66	89	115	181	261
	6	1	3	5	10	13	22	32	49	66	85	134	193
	4	1	2	4	7	9	16	23	35	48	61	97	140
	3	1	1	3	6	8	13	19	30	40	52	82	118
	2	1	1	3	5	7	11	16	25	34	44	69	99
XHH, XHHW, XHHW-2	1	1	1	1	3	5	8	12	19	25	32	51	74
	1/0	1	1	1	3	4	7	10	16	21	27	43	62
	2/0	0	1	1	2	3	6	8	13	17	23	36	52
	3/0	0	1	1	1	3	5	7	11	14	19	30	43
	4/0	0	1	1	1	2	4	6	9	12	15	24	35
	250	0	0	1	1	1	3	5	7	10	13	20	29
	300	0	0	1	1	1	3	4	6	8	11	17	25
	350	0	0	1	1	1	2	3	5	7	9	15	22
	400	0	0	0	1	1	1	3	5	6	8	13	19
	500	0	0	0	1	1	1	2	4	5	7	11	16
	600	0	0	0	1	1	1	1	3	4	5	9	13
	700	0	0	0	0	1	1	1	3	4	5	8	11
	750	0	0	0	0	1	1	1	2	3	4	7	11
	800	0	0	0	0	1	1	1	2	3	4	7	10
	900	0	0	0	0	1	1	1	2	3	4	6	9
	1000	0	0	0	0	0	1	1	1	3	3	6	8
	1250	0	0	0	0	0	1	1	1	1	3	4	6
	1500	0	0	0	0	0	1	1	1	1	2	4	5
	1750	0	0	0	0	0	0	1	1	1	1	3	5
	2000	0	0	0	0	0	0	1	1	1	1	3	4

Fixture Wires

Type	Conductor Size (AWG/kcmil)	Metric Designator (Trade Size) 16 (½)	21 (¾)	27 (1)	35 (1¼)	41 (1½)	53 (2)
FFH-2, RFH-2, RFHH-3	18	8	14	23	40	54	90
	16	6	12	19	33	46	76
SF-2, SFF-2	18	10	17	29	50	69	114
	16	8	14	24	42	57	94
	14	6	12	19	33	46	76
SF-1, SFF-1	18	17	31	51	89	122	202
RFHH-2, TF, TFF, XF, XFF RFH-1	18	13	23	38	66	90	149
RFFH-2, TF, TFF, XF, XFF	16	10	18	30	53	73	120
XF, XFF	14	8	14	24	42	57	94
TFN, TFFN	18	20	37	60	105	144	239
	16	16	28	46	80	110	183
PF, PFF, PGF, PGFF, PAF, PTF, PTFF, PAFF	18	19	35	57	100	137	227
	16	15	27	44	77	106	175
	14	11	20	33	58	79	131
HF, HFF, ZF, ZFF, ZHF	18	25	45	74	129	176	292
	16	18	33	54	95	130	216
	14	13	24	40	70	95	158
KF-2, KFF-2	18	36	65	107	187	256	424
	16	26	46	75	132	180	299
	14	17	31	52	90	124	205
	12	12	22	35	62	85	141
	10	8	14	24	42	57	94
KF-1, KFF-1	18	43	78	128	223	305	506
	16	30	55	90	157	214	355
	14	20	37	60	105	144	239
	12	13	24	40	70	95	158
	10	9	16	26	45	62	103
XF, XFF	12	4	8	12	22	30	50
	10	3	6	10	17	24	39

Note: This table is for concentric stranded conductors only. For compact stranded conductors, Table C10(A) should be used.

*Types RHH, RHW, and RHW-2 without outer covering.

Table C10(A) Maximum Number of Compact Conductors in Rigid PVC Conduit, Schedule 40 and HDPE Conduit (*Based on Table 1, Chapter 9*)

		Compact Conductors											
	Conductor	*Metric Designator (Trade Size)*											
Type	*Size (AWG/kcmil)*	*16 (½)*	*21 (¾)*	*27 (1)*	*35 (1¼)*	*41 (1½)*	*53 (2)*	*63 (2½)*	*78 (3)*	*91 (3½)*	*103 (4)*	*129 (5)*	*155 (6)*
THW, THW-2, THHW	8	1	4	6	11	15	26	37	57	76	98	155	224
	6	1	3	5	9	12	20	28	44	59	76	119	173
	4	1	1	3	6	9	15	21	33	44	57	89	129
	2	1	1	2	5	6	11	15	24	32	42	66	95
	1	1	1	1	3	4	7	11	17	23	29	46	67
	1/0	0	1	1	3	4	6	9	15	20	25	40	58
	2/0	0	1	1	2	3	5	8	12	16	21	34	49
	3/0	0	1	1	1	3	5	7	10	14	18	29	42
	4/0	0	1	1	1	2	4	5	9	12	15	24	35
	250	0	0	1	1	1	3	4	7	9	12	19	27
	300	0	0	1	1	1	2	4	6	8	10	16	24
	350	0	0	1	1	1	2	3	5	7	9	15	21
	400	0	0	0	1	1	1	3	5	6	8	13	19
	500	0	0	0	1	1	1	2	4	5	7	11	16
	600	0	0	0	1	1	1	1	3	4	5	9	13
	700	0	0	0	0	1	1	1	3	4	5	8	12
	750	0	0	0	0	1	1	1	2	3	5	7	11
	1000	0	0	0	0	1	1	1	1	3	4	6	9

Table C10(A) *Continued*

Compact Conductors

Type	Conductor Size (AWG/kcmil)	Metric Designator (Trade Size) 16 (½)	21 (¾)	27 (1)	35 (1¼)	41 (1½)	53 (2)	63 (2½)	78 (3)	91 (3½)	103 (4)	129 (5)	155 (6)
THHN, THWN, THWN-2	8	—	—	—	—	—	—	—	—	—	—	—	—
	6	2	4	7	13	17	29	41	64	86	111	175	253
	4	1	2	4	8	11	18	25	40	53	68	108	156
	2	1	1	3	5	8	13	18	28	38	49	77	112
	1	1	1	2	4	6	9	14	21	29	37	58	84
	1/0	1	1	1	3	5	8	12	18	24	31	49	72
	2/0	0	1	1	3	4	7	9	15	20	26	41	59
	3/0	0	1	1	2	3	5	8	12	17	22	34	50
	4/0	0	1	1	1	3	4	6	10	14	18	28	41
	250	0	0	1	1	1	3	5	8	11	14	22	32
	300	0	0	1	1	1	3	4	7	9	12	19	28
	350	0	0	1	1	1	3	4	6	8	10	17	24
	400	0	0	1	1	1	2	3	5	7	9	15	22
	500	0	0	0	1	1	1	3	4	6	8	13	18
	600	0	0	0	1	1	1	2	4	5	6	10	15
	700	0	0	0	1	1	1	1	3	4	5	9	13
	750	0	0	0	1	1	1	1	3	4	5	8	12
	1000	0	0	0	0	1	1	1	2	3	4	6	9

XHHW, XHHW-2	8	3	5	8	14	20	33	47	73	99	127	200	290
	6	1	4	6	11	15	25	35	55	73	94	149	215
	4	1	2	4	8	11	18	25	40	53	68	108	156
	2	1	1	3	5	8	13	18	28	38	49	77	112
	1	1	1	2	4	6	9	14	21	29	37	58	84
	1/0	1	1	1	3	5	8	12	18	24	31	49	72
	2/0	1	1	1	3	4	7	10	15	20	26	42	60
	3/0	0	1	1	2	3	5	8	12	17	22	34	50
	4/0	0	1	1	1	3	5	7	10	14	18	29	42
	250	0	0	1	1	1	4	5	8	11	14	23	33
	300	0	0	1	1	1	3	4	7	9	12	19	28
	350	0	0	1	1	1	3	4	6	8	11	17	25
	400	0	0	1	1	1	2	3	5	7	10	15	22
	500	0	0	0	1	1	1	3	4	6	8	13	18
	600	0	0	0	1	1	1	2	4	5	6	10	15
	700	0	0	0	1	1	1	1	3	4	5	9	13
	750	0	0	0	1	1	1	1	3	4	5	8	12
	1000	0	0	0	0	1	1	1	2	3	4	6	9

Definition: *Compact stranding* is the result of a manufacturing process where the standard conductor is compressed to the extent that the interstices (voids between strand wires) are virtually eliminated.

Table C11 Maximum Number of Conductors or Fixture Wires in Type A, Rigid PVC Conduit (*Based on Table 1, Chapter 9*)

Conductors

Type	Conductor Size (AWG/kcmil)	Metric Designator (Trade Size) 16 (½)	21 (¾)	27 (1)	35 (1¼)	41 (1½)	53 (2)	63 (2½)	78 (3)	91 (3½)	103 (4)
RHH,	14	5	9	15	24	31	49	74	112	146	187
RHW,	12	4	7	12	20	26	41	61	93	121	155
RHW-2	10	3	6	10	16	21	33	50	75	98	125
	8	1	3	5	8	11	17	26	39	51	65
	6	1	2	4	6	9	14	21	31	41	52
	4	1	1	3	5	7	11	16	24	32	41
	3	1	1	3	4	6	9	14	21	28	36
	2	1	1	2	4	5	8	12	18	24	31
	1	0	1	1	2	3	5	8	12	16	20
	1/0	0	1	1	2	3	5	7	10	14	18
	2/0	0	1	1	1	2	4	6	9	12	15
	3/0	0	1	1	1	1	3	5	8	10	13
	4/0	0	0	1	1	1	3	4	7	9	11
	250	0	0	1	1	1	1	3	5	7	8
	300	0	0	1	1	1	1	3	4	6	7
	350	0	0	0	1	1	1	2	4	5	7
	400	0	0	0	1	1	1	2	4	5	6
	500	0	0	0	1	1	1	1	3	4	5
	600	0	0	0	0	1	1	1	2	3	4
	700	0	0	0	0	1	1	1	2	3	4
	750	0	0	0	0	1	1	1	1	3	4
	800	0	0	0	0	1	1	1	1	3	3
	900	0	0	0	0	0	1	1	1	2	3
	1000	0	0	0	0	0	1	1	1	2	3
	1250	0	0	0	0	0	1	1	1	1	2
	1500	0	0	0	0	0	0	1	1	1	1
	1750	0	0	0	0	0	0	1	1	1	1
	2000	0	0	0	0	0	0	1	1	1	1
TW,	14	11	18	31	51	67	105	157	235	307	395
THHW,	12	8	14	24	39	51	80	120	181	236	303
THW,	10	6	10	18	29	38	60	89	135	176	226
THW-2	8	3	6	10	16	21	33	50	75	98	125
RHH*,	14	7	12	20	34	44	70	104	157	204	262
RHW*,	12	6	10	16	27	35	56	84	126	164	211
RHW-2*	10	4	8	13	21	28	44	65	98	128	165
	8	2	4	8	12	16	26	39	59	77	98
RHH, RHW*	6	1	3	6	9	13	20	30	45	59	75
TW,	4	1	2	4	7	9	15	22	33	44	56
THW,	3	1	1	4	6	8	13	19	29	37	48
THHW,	2	1	1	3	5	7	11	16	24	32	41
THW-2	1	1	1	1	3	5	7	11	17	22	29
	1/0	1	1	1	3	4	6	10	14	19	24
	2/0	0	1	1	2	3	5	8	12	16	21
	3/0	0	1	1	1	3	4	7	10	13	17
	4/0	0	1	1	1	2	4	6	9	11	14

Table C11 *Continued*

Conductors

Type	Conductor Size (AWG/kcmil)	*Metric Designator (Trade Size)* 16 (½)	21 (¾)	27 (1)	35 (1¼)	41 (1½)	53 (2)	63 (2½)	78 (3)	91 (3½)	103 (4)
TW, THW, THHW, THW-2	250	0	0	1	1	1	3	4	7	9	12
	300	0	0	1	1	1	2	4	6	8	10
	350	0	0	1	1	1	2	3	5	7	9
	400	0	0	1	1	1	1	3	5	6	8
	500	0	0	0	1	1	1	2	4	5	7
	600	0	0	0	1	1	1	1	3	4	5
	700	0	0	0	1	1	1	1	3	4	5
	750	0	0	0	1	1	1	1	3	3	4
	800	0	0	0	0	1	1	1	2	3	4
	900	0	0	0	0	1	1	1	2	3	4
	1000	0	0	0	0	1	1	1	1	3	3
	1250	0	0	0	0	0	1	1	1	1	3
	1500	0	0	0	0	0	1	1	1	1	2
	1750	0	0	0	0	0	0	1	1	1	1
	2000	0	0	0	0	0	0	1	1	1	1
THHN, THWN, THWN-2	14	16	27	44	73	96	150	225	338	441	566
	12	11	19	32	53	70	109	164	246	321	412
	10	7	12	20	33	44	69	103	155	202	260
	8	4	7	12	19	25	40	59	89	117	150
	6	3	5	8	14	18	28	43	64	84	108
	4	1	3	5	8	11	17	26	39	52	66
	3	1	2	4	7	9	15	22	33	44	56
	2	1	1	3	6	8	12	19	28	37	47
	1	1	1	2	4	6	9	14	21	27	35
	1/0	1	1	2	4	5	8	11	17	23	29
	2/0	1	1	1	3	4	6	10	14	19	24
	3/0	0	1	1	2	3	5	8	12	16	20
	4/0	0	1	1	1	3	4	6	10	13	17
	250	0	1	1	1	2	3	5	8	10	14
	300	0	0	1	1	1	3	4	7	9	12
	350	0	0	1	1	1	2	4	6	8	10
	400	0	0	1	1	1	2	3	5	7	9
	500	0	0	1	1	1	1	3	4	6	7
	600	0	0	0	1	1	1	2	3	5	6
	700	0	0	0	1	1	1	1	3	4	5
	750	0	0	0	1	1	1	1	3	4	5
	800	0	0	0	1	1	1	1	3	4	5
	900	0	0	0	0	1	1	1	2	3	4
	1000	0	0	0	0	1	1	1	2	3	4
FEP, FEPB, PFA, PFAH, TFE	14	15	26	43	70	93	146	218	327	427	549
	12	11	19	31	51	68	106	159	239	312	400
	10	8	13	22	37	48	76	114	171	224	287
	8	4	8	13	21	28	44	65	98	128	165
	6	3	5	9	15	20	31	46	70	91	117
	4	1	4	6	10	14	21	32	49	64	82
	3	1	3	5	8	11	18	27	40	53	68
	2	1	2	4	7	9	15	22	33	44	56

Table C11 *Continued*

Type	Conductor Size (AWG/kcmil)	Conductors: Metric Designator (Trade Size) 16 (½)	21 (¾)	27 (1)	35 (1¼)	41 (1½)	53 (2)	63 (2½)	78 (3)	91 (3½)	103 (4)
PFA, PFAH, TFE	1	1	1	3	5	6	10	15	23	30	39
PFA, PFAH, TFE, Z	1/0	1	1	2	4	5	8	13	19	25	32
	2/0	1	1	1	3	4	7	10	16	21	27
	3/0	1	1	1	3	3	6	9	13	17	22
	4/0	0	1	1	2	3	5	7	11	14	18
Z	14	18	31	52	85	112	175	263	395	515	661
	12	13	22	37	60	79	124	186	280	365	469
	10	8	13	22	37	48	76	114	171	224	287
	8	5	8	14	23	30	48	72	108	141	181
	6	3	6	10	16	21	34	50	76	99	127
	4	2	4	7	11	15	23	35	52	68	88
	3	1	3	5	8	11	17	25	38	50	64
	2	1	2	4	7	9	14	21	32	41	53
	1	1	1	3	5	7	11	17	26	33	43
XHH, XHHW, XHHW-2, ZW	14	11	18	31	51	67	105	157	235	307	395
	12	8	14	24	39	51	80	120	181	236	303
	10	6	10	18	29	38	60	89	135	176	226
	8	3	6	10	16	21	33	50	75	98	125
	6	2	4	7	12	15	24	37	55	72	93
	4	1	3	5	8	11	18	26	40	52	67
	3	1	2	4	7	9	15	22	34	44	57
	2	1	1	3	6	8	12	19	28	37	48
XHH, XHHW, XHHW-2	1	1	1	3	4	6	9	14	21	28	35
	1/0	1	1	2	4	5	8	12	18	23	30
	2/0	1	1	1	3	4	6	10	15	19	25
	3/0	0	1	1	2	3	5	8	12	16	20
	4/0	0	1	1	1	3	4	7	10	13	17
	250	0	1	1	1	2	3	5	8	11	14
	300	0	0	1	1	1	3	5	7	9	12
	350	0	0	1	1	1	3	4	6	8	10
	400	0	0	1	1	1	2	3	5	7	9
	500	0	0	1	1	1	1	3	4	6	8
	600	0	0	0	1	1	1	2	3	5	6
	700	0	0	0	1	1	1	1	3	4	5
	750	0	0	0	1	1	1	1	3	4	5
	800	0	0	0	1	1	1	1	3	4	5
	900	0	0	0	0	1	1	1	2	3	4
	1000	0	0	0	0	1	1	1	2	3	4
	1250	0	0	0	0	0	1	1	1	2	3
	1500	0	0	0	0	0	1	1	1	1	2
	1750	0	0	0	0	0	1	1	1	1	2
	2000	0	0	0	0	0	0	1	1	1	1

Fixture Wires

Type	Conductor Size (AWG/kcmil)	*Metric Designator (Trade Size)* 16 (½)	21 (¾)	27 (1)	35 (1¼)	41 (1½)	53 (2)
FFH-2, RFH-2, RFHH-3	18	10	18	30	48	64	100
	16	9	15	25	41	54	85
SF-2, SFF-2	18	13	22	37	61	81	127
	16	11	18	31	51	67	105
	14	9	15	25	41	54	85
SF-1, SFF-1	18	23	40	66	108	143	224
RFH-1, RFHH-2, TF, TFF, XF, XFF	18	17	29	49	80	105	165
RFHH-2, TF, TFF, XF, XFF	16	14	24	39	65	85	134
XF, XFF	14	11	18	31	51	67	105
TFN, TFFN	18	28	47	79	128	169	265
	16	21	36	60	98	129	202
PF, PFF, PGF, PGFF, PAF, PTF, PTFF, PAFF	18	26	45	74	122	160	251
	16	20	34	58	94	124	194
	14	15	26	43	70	93	146
HF, HFF, ZF, ZFF, ZHF	18	34	58	96	157	206	324
	16	25	42	71	116	152	239
	14	18	31	52	85	112	175
KF-2, KFF-2	18	49	84	140	228	300	470
	16	35	59	98	160	211	331
	14	24	40	67	110	145	228
	12	16	28	46	76	100	157
	10	11	18	31	51	67	105
KF-1, KFF-1	18	59	100	167	272	357	561
	16	41	70	117	191	251	394
	14	28	47	79	128	169	265
	12	18	31	52	85	112	175
	10	12	20	34	55	73	115
XF, XFF	12	6	10	16	27	35	56
	10	4	8	13	21	28	44

Note: This table is for concentric stranded conductors only. For compact stranded conductors, Table C11(A) should be used.

*Types RHH, RHW, and RHW-2 without outer covering.

Table C11(A) Maximum Number of Compact Conductors in Type A, Rigid PVC Conduit (*Based on Table 1, Chapter 9*)

Compact Conductors

Type	Conductor Size (AWG/kcmil)	*Metric Designator (Trade Size)* 16 (½)	21 (¾)	27 (1)	35 (1¼)	41 (1½)	53 (2)	63 (2½)	78 (3)	91 (3½)	103 (4)
THW, THW-2, THHW	8	3	5	8	14	18	28	42	64	84	107
	6	2	4	6	10	14	22	33	49	65	83
	4	1	3	5	8	10	16	24	37	48	62
	2	1	1	3	6	7	12	18	27	36	46
	1	1	1	2	4	5	8	13	19	25	32
	1/0	1	1	1	3	4	7	11	16	21	28
	2/0	1	1	1	3	4	6	9	14	18	23
	3/0	0	1	1	2	3	5	8	12	15	20
	4/0	0	1	1	1	3	4	6	10	13	17
	250	0	1	1	1	1	3	5	8	10	13
	300	0	0	1	1	1	3	4	7	9	11
	350	0	0	1	1	1	2	4	6	8	10
	400	0	0	1	1	1	2	3	5	7	9
	500	0	0	1	1	1	1	3	4	6	8
	600	0	0	0	1	1	1	2	3	5	6
	700	0	0	0	1	1	1	1	3	4	5
	750	0	0	0	1	1	1	1	3	4	5
	1000	0	0	0	0	1	1	1	2	3	4
THHN, THWN, THWN-2	8	—	—	—	—	—	—	—	—	—	—
	6	3	5	9	15	20	32	48	72	94	121
	4	1	3	6	9	12	20	30	45	58	75
	2	1	2	4	7	9	14	21	32	42	54
	1	1	1	3	5	7	10	16	24	31	40
	1/0	1	1	2	4	6	9	13	20	27	34
	2/0	1	1	1	3	5	7	11	17	22	28
	3/0	1	1	1	3	4	6	9	14	18	24
	4/0	0	1	1	2	3	5	8	11	15	19
	250	0	1	1	1	2	4	6	9	12	15
	300	0	1	1	1	1	3	5	8	10	13
	350	0	0	1	1	1	3	4	7	9	11
	400	0	0	1	1	1	2	4	6	8	10
	500	0	0	1	1	1	2	3	5	7	9
	600	0	0	0	1	1	1	3	4	5	7
	700	0	0	0	1	1	1	2	3	5	6
	750	0	0	0	1	1	1	2	3	4	6
	1000	0	0	0	0	1	1	1	2	3	4
XHHW, XHHW-2	8	4	6	11	18	23	37	55	83	108	139
	6	3	5	8	13	17	27	41	62	80	103
	4	1	3	6	9	12	20	30	45	58	75
	2	1	2	4	7	9	14	21	32	42	54
	1	1	1	3	5	7	10	16	24	31	40

Table C11(A) *Continued*

Compact Conductors

Type	Conductor Size (AWG/kcmil)	Metric Designator (Trade Size) 16 (½)	21 (¾)	27 (1)	35 (1¼)	41 (1½)	53 (2)	63 (2½)	78 (3)	91 (3½)	103 (4)
XHHW, XHHW-2	1/0	1	1	2	4	6	9	13	20	27	34
	2/0	1	1	1	3	5	7	11	17	22	29
	3/0	1	1	1	3	4	6	9	14	18	24
	4/0	0	1	1	2	3	5	8	12	15	20
	250	0	1	1	1	2	4	6	9	12	16
	300	0	1	1	1	1	3	5	8	10	13
	350	0	0	1	1	1	3	5	7	9	12
	400	0	0	1	1	1	3	4	6	8	11
	500	0	0	1	1	1	2	3	5	7	9
	600	0	0	0	1	1	1	3	4	5	7
	700	0	0	0	1	1	1	2	3	5	6
	750	0	0	0	1	1	1	2	3	4	6
	1000	0	0	0	0	1	1	1	2	3	4

Definition: *Compact stranding* is the result of a manufacturing process where the standard conductor is compressed to the extent that the interstices (voids between strand wires) are virtually eliminated.

Table C12 Maximum of Conductors in Type EB, PVC Conduit (*Based on Table 1, Chapter 9*)

		Conductors					
		Metric Designator (Trade Size)					
Type	*Conductor Size (AWG/kcmil)*	*53 (2)*	*78 (3)*	*91 (3½)*	*103 (4)*	*129 (5)*	*155 (6)*
RHH, RHW, RHW-2	14	53	119	155	197	303	430
	12	44	98	128	163	251	357
RHH, RHW, RHW-2	10	35	79	104	132	203	288
	8	18	41	54	69	106	151
	6	15	33	43	55	85	121
	4	11	26	34	43	66	94
	3	10	23	30	38	58	83
	2	9	20	26	33	50	72
	1	6	13	17	21	33	47
	1/0	5	11	15	19	29	41
	2/0	4	10	13	16	25	36
	3/0	4	8	11	14	22	31
	4/0	3	7	9	12	18	26
	250	2	5	7	9	14	20
	300	1	5	6	8	12	17
	350	1	4	5	7	11	16
	400	1	4	5	6	10	14
	500	1	3	4	5	9	12
	600	1	3	3	4	7	10
	700	1	2	3	4	6	9
	750	1	2	3	4	6	9
	800	1	2	3	4	6	8
	900	1	1	2	3	5	7
	1000	1	1	2	3	5	7
	1250	1	1	1	2	3	5
	1500	0	1	1	1	3	4
	1750	0	1	1	1	3	4
	2000	0	1	1	1	2	3
TW, THHW, THW, THW-2	14	111	250	327	415	638	907
	12	85	192	251	319	490	696
	10	63	143	187	238	365	519
	8	35	79	104	132	203	288
RHH*, RHW*, WH-2*	14	74	166	217	276	424	603
RHH*, HW*, RHW-2*	12	59	134	175	222	341	485
	10	46	104	136	173	266	378
RHH*, HW*, RHW-2*	8	28	62	81	104	159	227

Table C12 *Continued*

Type	Conductor Size (AWG/kcmil)	Conductors — Metric Designator (Trade Size) 53 (2)	78 (3)	91 (3½)	103 (4)	129 (5)	155 (6)
RHH*, RHW*, RHW-2*, TW, THW, THHW, THW-2	6	21	48	62	79	122	173
	4	16	36	46	59	91	129
	3	13	30	40	51	78	111
	2	11	26	34	43	66	94
	1	8	18	24	30	46	66
	1/0	7	15	20	26	40	56
	2/0	6	13	17	22	34	48
	3/0	5	11	14	18	28	40
	4/0	4	9	12	15	24	34
	250	3	7	10	12	19	27
	300	3	6	8	11	17	24
	350	2	6	7	9	15	21
	400	2	5	7	8	13	19
	500	1	4	5	7	11	16
	600	1	3	4	6	9	13
	700	1	3	4	5	8	11
	750	1	3	4	5	7	11
	800	1	3	3	4	7	10
	900	1	2	3	4	6	9
	1000	1	2	3	4	6	8
	1250	1	1	2	3	4	6
	1500	1	1	1	2	4	6
	1750	1	1	1	2	3	5
	2000	0	1	1	1	3	4
THHN, THWN, THWN-2	14	159	359	468	595	915	1300
	12	116	262	342	434	667	948
	10	73	165	215	274	420	597
	8	42	95	124	158	242	344
	6	30	68	89	114	175	248
	4	19	42	55	70	107	153
	3	16	36	46	59	91	129
	2	13	30	39	50	76	109
	1	10	22	29	37	57	80
	1/0	8	18	24	31	48	68
	2/0	7	15	20	26	40	56
	3/0	5	13	17	21	33	47
	4/0	4	10	14	18	27	39
	250	4	8	11	14	22	31
	300	3	7	10	12	19	27
	350	3	6	8	11	17	24
	400	2	6	7	10	15	21
	500	1	5	6	8	12	18

Table C12 *Continued*

		Conductors					
		Metric Designator (Trade Size)					
Type	*Conductor Size (AWG/kcmil)*	*53 (2)*	*78 (3)*	*91 (3½)*	*103 (4)*	*129 (5)*	*155 (6)*
THHN, THWN, THWN-2	600	1	4	5	6	10	14
	700	1	3	4	6	9	12
	750	1	3	4	5	8	12
	800	1	3	4	5	8	11
	900	1	3	3	4	7	10
	1000	1	2	3	4	6	9
FEP, FEPB, PFA, PFAH, TFE	14	155	348	454	578	888	1261
	12	113	254	332	422	648	920
	10	81	182	238	302	465	660
	8	46	104	136	173	266	378
	6	33	74	97	123	189	269
	4	23	52	68	86	132	188
	3	19	43	56	72	110	157
	2	16	36	46	59	91	129
PFA, PFAH, TFE	1	11	25	32	41	63	90
PFA, PFAH, TFE, Z	1/0	9	20	27	34	53	75
	2/0	7	17	22	28	43	62
	3/0	6	14	18	23	36	51
	4/0	5	11	15	19	29	42
Z	14	186	419	547	696	1069	1519
	12	132	297	388	494	759	1078
	10	81	182	238	302	465	660
	8	51	115	150	191	294	417
	6	36	81	105	134	206	293
	4	24	55	72	92	142	201
	3	18	40	53	67	104	147
	2	15	34	44	56	86	122
	1	12	27	36	45	70	99
XHH, XHHW, XHHW-2, ZW	14	111	250	327	415	638	907
	12	85	192	251	319	490	696
	10	63	143	187	238	365	519
	8	35	79	104	132	203	288
	6	26	59	77	98	150	213
	4	19	42	56	71	109	155
	3	16	36	47	60	92	131
	2	13	30	39	50	77	110

Table C12 *Continued*

		Conductors					
		Metric Designator (Trade Size)					
Type	*Conductor Size (AWG/kcmil)*	*53 (2)*	*78 (3)*	*91 (3½)*	*103 (4)*	*129 (5)*	*155 (6)*
XHH, XHHW, XHHW-2	1	10	22	29	37	58	82
	1/0	8	19	25	31	48	69
	2/0	7	16	20	26	40	57
	3/0	6	13	17	22	33	47
	4/0	5	11	14	18	27	39
	250	4	9	11	15	22	32
	300	3	7	10	12	19	28
	350	3	6	9	11	17	24
	400	2	6	8	10	15	22
	500	1	5	6	8	12	18
	600	1	4	5	6	10	14
	700	1	3	4	6	9	12
	750	1	3	4	5	8	12
	800	1	3	4	5	8	11
	900	1	3	3	4	7	10
	1000	1	2	3	4	6	9
	1250	1	1	2	3	5	7
	1500	1	1	1	3	4	6
	1750	1	1	1	2	4	5
	2000	0	1	1	1	3	5

Note: This table is for concentric stranded conductors only. For compact stranded conductors, Table C12(A) should be used.

*Types RHH, RHW, and RHW-2 without outer covering.

Table C12(A) Maximum Number of Compact Conductors in Type EB, PVC Conduit (*Based on Table 1, Chapter 9*)

Compact Conductors

		Metric Designator (Trade Size)					
Type	*Conductor Size (AWG/kcmil)*	*53 (2)*	*78 (3)*	*91 (3½)*	*103 (4)*	*129 (5)*	*155 (6)*
THW, THW-2, THHW	8	30	68	89	113	174	247
	6	23	52	69	87	134	191
	4	17	39	51	65	100	143
	2	13	29	38	48	74	105
	1	9	20	26	34	52	74
	1/0	8	17	23	29	45	64
	2/0	6	15	19	24	38	54
	3/0	5	12	16	21	32	46
	4/0	4	10	14	17	27	38
	250	3	8	11	14	21	30
	300	3	7	9	12	19	26
	350	3	6	8	11	17	24
	400	2	6	7	10	15	21
	500	1	5	6	8	12	18
	600	1	4	5	6	10	14
	700	1	3	4	6	9	13
	750	1	3	4	5	8	12
	1000	1	2	3	4	7	9
THHN, THWN, THWN-2	8	—	—	—	—	—	—
	6	34	77	100	128	196	279
	4	21	47	62	79	121	172
	2	15	34	44	57	87	124
	1	11	25	33	42	65	93
	1/0	9	22	28	36	56	79
	2/0	8	18	23	30	46	65
	3/0	6	15	20	25	38	55
	4/0	5	12	16	20	32	45
	250	4	10	13	16	25	35
	300	4	8	11	14	22	31
	350	3	7	9	12	19	27
	400	3	6	8	11	17	24
	500	2	5	7	9	14	20
	600	1	4	6	7	11	16
	700	1	4	5	6	10	14
	750	1	4	5	6	9	14
	1000	1	3	3	4	7	10

Table C12(A) *Continued*

Compact Conductors

Type	Conductor Size (AWG/kcmil)	53 (2)	78 (3)	91 (3½)	103 (4)	129 (5)	155 (6)
		Metric Designator (Trade Size)					
XHHW, XHHW-2	8	39	88	115	146	225	320
	6	29	65	85	109	167	238
	4	21	47	62	79	121	172
	2	15	34	44	57	87	124
	1	11	25	33	42	65	93
	1/0	9	22	28	36	56	79
	2/0	8	18	24	30	47	67
	3/0	6	15	20	25	38	55
	4/0	5	12	16	21	32	46
	250	4	10	13	17	26	37
	300	4	8	11	14	22	31
	350	3	7	10	12	19	28
	400	3	7	9	11	17	25
	500	2	5	7	9	14	20
	600	1	4	6	7	11	16
	700	1	4	5	6	10	14
	750	1	3	5	6	9	13
	1000	1	3	4	5	7	10

Definition: *Compact stranding* is the result of a manufacturing process where the standard conductor is compressed to the extent that the interstices (voids between strand wires) are virtually eliminated.

Annex D

EXAMPLES

Selection of Conductors. In the following examples, the results are generally expressed in amperes (A). To select conductor sizes, refer to the 0 through 2000 Volt (V) ampacity tables of Article 310 and the rules of 310.15 that pertain to these tables.

Voltage. For uniform application of Articles 210, 215, and 220, a nominal voltage of 120, 120/240, 240, and 208Y/120 volts shall be used in computing the ampere load on the conductor.

Fractions of an Ampere. Except where the computations result in a major fraction of an ampere (0.5 or larger), such fractions are permitted to be dropped.

Power Factor. Calculations in the following examples are based, for convenience, on the assumption that all loads have the same power factor (PF).

Ranges. For the computation of the range loads in these examples, Column A of Table 220.19 has been used. For optional methods, see Columns B and C of Table 220.19. Except where the computations result in a major fraction of a kilowatt (0.5 or larger), such fractions are permitted to be dropped.

SI Units. For metric conversions, $0.093\ m^2 = 1\ ft^2$ and $0.3048\ m = 1\ ft$.

Example No. D1(a) One-Family Dwelling

The dwelling has a floor area of 1500 square feet exclusive of an unfinished cellar not adaptable for future use, unfinished attic, and open porches. Appliances are a 12-kW range and a 5.5-kW, 240-volt dryer. Assume range and dryer kW ratings equivalent to kVA ratings in accordance with Sections 220.18 and 220.19.

Computed Load [See 220.10]

General lighting load: 1500 sq. ft. at 3 volt-amperes per sq. ft. = 4500 volt-amperes.

Minimum Number of Branch Circuits Required [See Section 210.11(A)]

General lighting load: 4500 volt-amperes ÷ 120 V = 37.5 A. This requires three 15-A, 2-wire or two 20-A, 2-wire circuits.

Small appliance load: two 2-wire, 20-A circuits [see 210.11(C)(1)].

Laundry load: one 2-wire 20-A circuit [see 210.11(C)(2)].

Bathroom branch circuit: One 2-wire, 20-A circuit (no additional calculation is required for this circuit) [see 210.11(C)(3)].

Minimum Size Feeder Required [See 220.10]

General lighting	4500 VA
Small appliance load	3000 VA
Laundry	1500 VA
Total general light and small appliance	9000 VA
3000 volt-amperes at 100%	3000 VA
9000 VA – 3000 VA = 6000 volt-amperes at 35%	2100 VA
Net load	5100 VA
Range load (see Table 220.19)	8000 VA
Dryer load (see Table 220.18)	5500 VA
Net computed load	18,600 VA

Net computed load for 120/240-volt 3-wire single-phase service or feeder

$$18{,}600 \text{ VA} \div 240 \text{ V} = 77.5 \text{ A}$$

Sections 230.42(B) and 230.79 require service conductors and disconnecting means rated not less than 100 amperes.

Calculation for Neutral for Feeder and Service

Lighting and small appliance load	5100 VA
Range load, 8000 volt-amperes at 70% (see 220.22)	5600 VA
Dryer: 5500 volt-amperes at 70% (see 220.22)	3850 VA
Total	14,550 VA

Computed Load for Neutral

$$14{,}550 \text{ VA} \div 240 \text{ V} = 60.6 \text{ A}$$

Example No. D1(b) One-Family Dwelling

Assume same conditions as Example No. D1(a), plus addition of one 6-ampere, 230-volt room air-conditioning unit and one 12-ampere, 115-volt room air-conditioning unit,* one 8-ampere, 115-volt rated waste disposer, and one 10-ampere, 120-volt rated dishwasher. See Article 430 for general motors and Article 440, Part VII, for air-conditioning equipment. Motors have nameplate ratings of 115 V and 230 V for use on 120-V and 240-V nominal voltage systems.

From Example No. D1(a), feeder current is 78 amperes (3-wire 240 volts).

	Line A	*Neutral*	*Line B*
Amperes from Example No. D1(a)	78	61	78
One 230-V air conditioner	6	–	6
One 115-V air conditioner and 120-V dishwasher	12	12	10
One 115-V disposal	–	8	8
25% of largest motor (see 430.24)	3	2	2
Total amperes per line	99	83	104

Therefore, the service would be rated 110 A.

*(For feeder neutral, use largest of the two appliances for unbalance.)

Example No. D2(a) Optional Calculation for One-Family Dwelling Heating Larger than Air Conditioning [See Section 220.30]

The dwelling has a floor area of 1500 square feet exclusive of an unfinished cellar not adaptable for future use, unfinished attic, and open porches. It has a 12-kW range, a 2.5-kW water heater, a 1.2-kW dishwasher, 9 kW of electric space heating installed in five rooms, a 5-kW clothes dryer, and a 6-ampere, 230-volt room air-conditioning unit. Assume range, water heater, dishwasher, space heating, and clothes dryer kW ratings equivalent to kVA.

Air conditioner kVA calculation is 6A × 230V ÷ 1000 = 1.38 kVA

This 1.38 kVA [Item 1 from 220.30(C)] is less than 40% of 9 kVA of separately controlled electric heat [Item 6 from 220.30(C)], so the 1.38 kVA need not be included in the service calculation.

General Load

1500 sq. ft. at 3 volt-amperes	4500 VA
Two 20-amp. appliance outlet circuits at 1500 volt-amperes each	3000 VA
Laundry circuit	1500 VA
Range (at nameplate rating)	12,000 VA
Water heater	2500 VA
Dishwasher	1200 VA
Clothes dryer	5000 VA
Total	29,700 VA

Application of Demand Factor [See 220.30(B)]

First 10 kVA of general load at 100%	=	10,000 VA
Remainder of general load at 40% (19.7 kVA × .4)	=	7880 VA
Total of general load	=	17,880 VA
9 kVA of heat at 40% (9 × .4)	=	3600 VA
Total	=	21,480 VA

Calculated load for service size

21,480 VA ÷ 240 volts = 89.5 amperes

Therefore, the minimum service size would be 100 amperes in accordance with 230.42 and 230.79.

Feeder Neutral Load (per 220.22)

1500 sq. ft. at 3 volt-amperes	4500 VA
Three 20-amp. circuits at 1500 volt-amperes	4500 VA
Total	9000 VA
3000 volt-amperes at 100%	3000 VA
9000 VA – 3000 VA = 6000 volt-amperes at 35%	2100 VA
Subtotal	5100 VA
Range—8 kVA at 70%	5600 VA
Clothes dryer—5 kVA at 70%	3500 VA
Dishwasher	1200 VA
Total	15,400 VA

Calculated Load for Neutral

15,400 VA ÷ 240 V = 64.2 A

Example No. D2(b) Optional Calculation for One-Family Dwelling, Air Conditioning Larger than Heating [See 220.30(A) and 220.30(C)]

The dwelling has a floor area of 1500 square feet exclusive of an unfinished cellar not adaptable for future use, unfinished attic, and open porches. It has two 20-ampere small appliance circuits, one 20-ampere laundry circuit, two 4-kW wall-mounted ovens, one 5.1-kW counter-mounted cooking unit, a 4.5-kW water heater, a 1.2-kW dishwasher, a 5-kW combination clothes washer and dryer, six 7-ampere, 230-volt room air-conditioning units, and a 1.5-kW permanently installed bathroom space heater. Assume wall-mounted ovens, counter-mounted cooking unit, water heater, dishwasher, and combination clothes washer and dryer kW ratings equivalent to kVA.

Air Conditioning kVA Calculation

Total amperes: 6 units × 7 A = 42 A

42 A × 240 V ÷ 1000 = 10.08 kVA (assume P.F. = 1.0).

Load Included at 100%

Air conditioning: Included below [See Item 1 in 220.30(C)]
Space heater: Omit [See Item 5 in 220.30(C)]

General Load

1500 sq. ft. at 3 volt-amperes	4500 VA	
Two 20-amp. small appliance circuits at 1500 volt-amperes each	3000 VA	
Laundry circuit	1500 VA	
2 ovens	8000 VA	
1 cooking unit	5100 VA	
Water heater	4500 VA	
Dishwasher	1200 VA	
Washer/dryer	5000 VA	
Total general load	32,800 VA	
1st 10 kVA at 100%		10,000 VA
Remainder at 40% (22.8 kVA × .4 × 1000)		9,120 VA
Subtotal general load		19,120 VA
Air conditioning		10,080 VA
		29,200 VA

Calculated Load for Service

29,200 VA ÷ 240 V = 122 A (service rating)

Feeder Neutral Load (per 220.22)

Assume that the two 4-kVA wall-mounted ovens are supplied by one branch circuit, the 5.1-kVA counter-mounted cooking unit by a separate circuit.

1500 sq. ft. at 3 volt-amperes	4500 VA
Three 20-amp circuits at 1500 volt-amperes	4500 VA
Subtotal	9000 VA
3000 volt-amperes at 100%	3000 VA
9000 VA – 3000 VA = 6000 volt-amperes at 35%	2100 VA
Subtotal	5100 VA

Two 4-kVA ovens plus one 5.1-kVA cooking unit = 13.1 kVA. Table 220.19 permits 55% demand factor or

$$13.1 \text{ kVA} \times 0.55 = 7.2\text{-kVA feeder capacity}$$

Subtotal from above	5100 VA
Ovens and cooking unit: 7200 VA × 70% for neutral load	5040 VA
Clothes washer/dryer – 5 kVA × 70% for neutral load	3500 VA
Dishwasher	1200 VA
Total	14,840 VA

Calculated Load for Neutral

$$14{,}840 \text{ VA} \div 240 \text{ V} = 61.83 \text{ (use 62 A)}$$

Example No. D2(c) Optional Calculation for One-Family Dwelling with Heat Pump (Single-Phase, 240/120-Volt Service) (See 220.30)

Dwelling has a floor area of 2000 sq. ft. exclusive of an unfinished cellar not adaptable for future use, unfinished attic, and open porches. It has a 12-kW range, 4.5-kW water heater, a 1.2-kW dishwasher, a 5-kW clothes dryer, a 2½-ton (24-ampere) heat pump with 15-kW of back-up heat.

Heat Pump kVA Calculation

$$24\text{A} \times 240\text{V} \div 1000 = 5.76 \text{ kVA}$$

This 5.76 kVA is less than 15 kVA of the backup heat; therefore, the heat pump load need not be included in the service calculation. [See 220.30(C).]

General Load

2000 sq. ft. at 3 volt-amperes	6000 VA
Two 20-ampere appliance outlet circuits at 1500 volt-amperes each	3000 VA
Laundry circuit	1500 VA
Range (at nameplate rating)	12,000 VA
Water heater	4500 VA
Dishwasher	1200 VA
Clothes dryer	5000 VA
Subtotal general load	33,200 VA
First 10 kVA of general load at 100% =	10,000 VA
Remainder of general load at 40% (23,200 VA × 0.4) =	9280 VA
Total net general load =	19,280 VA

Heat Pump and Supplementary Heat*

240V × 24A = 5,760 VA

15-kW electric heat: 5760 VA + 15,000 VA = 20,760 VA or 20.76 kVA

20.76 kVA × 100% = 20.76 kVA

* If supplementary heat is not on at the same time as heat pump, heat pump kVA need not be added to total.

Totals:

Net general load	19,280 VA
Heat pump and supplementary heat	20,760 VA
Total	40,040 VA

Calculated Load for Service

40.04 kVA × 1000 ÷ 240 V = 166.8 A

Therefore, this dwelling unit may be served by a 175-ampere service.

Example No. D3 Store Building

A store 50 ft. by 60 ft., or 3000 sq. ft., has 30 ft. of show window. There are a total of 80 duplex receptacles. The service is 120/240-volt, single-phase 3-wire service. Actual connected lighting load, 8500 volt-amperes.

Computed Load (See 220.10)

Noncontinuous Loads

Receptacle Load (see 220.13)

80 receptacles at 180 VA	14,400 volt-amperes
10,000 VA at 100%	10,000 volt-amperes
14,400 VA – 10,000 VA = 4400 VA at 50%	2200 volt-amperes
Subtotal	12,200 volt-amperes

Continuous Loads

General Lighting*	
3000 sq. ft. at 3 volt-amperes per sq. ft.	9000 volt-amperes
Show Window Lighting Load	
30 ft. at 200 volt-amperes per foot	6000 volt-amperes
Outside sign circuit 1200 volt-amperes [see 220.3(B)(6)]	1200 volt-amperes
Subtotal	16,200 volt-amperes
Subtotal from noncontinuous	12,200 volt-amperes
Subtotal noncontinuous loads plus continuous loads	28,400 volt-amperes

*In the example, 125 percent of the actual connected lighting load (8500 VA × 1.25 = 10,625 VA) is less than 125 percent of the load from Table 220.3(A), so the minimum lighting load from Table 220.3(A) is used in the calculation. Had the actual lighting load been greater than the value computed from Table 220.3(A), 125 percent of the actual connected lighting load would have been used.

Minimum Number of Branch Circuits Required

General Lighting: Branch circuits need only be installed to supply the actual connected load [see 221.11(B)].

8500 VA × 1.25 = 10,625 volt-amperes
10,625 volt-amperes ÷ 240 volts =
44 amperes for 3-wire, 120/240 V

The lighting load would be permitted to be served by 2-wire or 3-wire, 15- or 20-ampere circuits with combined capacity equal to 44 amperes or greater for 3-wire circuits or 88 amperes or greater for 2-wire circuits. The feeder capacity as well as the number of branch-circuit positions available for lighting circuits in the panelboard must reflect the full calculated load of 9000 VA × 1.25 = 11,250 volt-amperes.

Show Window

6,000 VA × 1.25 = 7500 VA

7500 volt-amperes ÷ 240 volts

= 31 amperes for 3-wire, 120/240V

The show window lighting is permitted to be served by 2-wire or 3-wire circuits with a capacity equal to 31 amperes or greater for 3-wire circuits or 62 amperes or greater for 2-wire circuits.

Receptacles required by 210.62 are assumed to be included in the receptacle load above if these receptacles do not supply the show window lighting load.

Receptacles

Receptacle Load:

14,400 volt-amperes ÷ 240 volts

= 60 amperes for 3-wire, 120/240 V

The receptacle load would be permitted to be served by 2-wire or 3-wire circuits with a capacity equal to 60 amperes or greater for 3-wire circuits or 120 amperes or greater for 2-wire circuits.

Minimum Size Feeder (or Service) Overcurrent Protection [See 215.3 or Section 230.90]

Subtotal noncontinuous loads	12,200 volt-amperes
Subtotal continuous load @ 125%	
16,200 VA × 1.25	20,250 volt-amperes
Total load	32,450 volt-amperes

32,450 VA ÷ 240 volts = 135 amperes

The next higher standard size is 150 amperes (See 240.6).

Minimum Size Feeder (or Service Conductors) Required [See 215.2, 230.42(A)]

For 120/240-volt, 3-wire system: 32,450 volt-amperes ÷ 240 volts = 135 amperes.

Service or feeder conductor is 1/0 Cu per 215.3 and Table 310.16 (with 75°C terminations).

Example No. D4(a) Multifamily Dwelling

Multifamily dwelling has 40 dwelling units.

Meters in two banks of 20 each and individual feeders to each dwelling unit.

One-half of the dwelling units are equipped with electric ranges not exceeding 12 kW each. Assume range kW rating equivalent to kVA rating in accordance with 220.19. Other half of ranges are gas ranges.

Area of each dwelling unit is 840 square feet.

Laundry facilities on premises are available to all tenants. Add no circuit to individual dwelling unit. Add no circuit to individual dwelling unit.

Computed Load for Each Dwelling Unit (See Article 220)

General lighting: 840 sq. ft. at 3 volt-amperes per sq. ft.	2520 VA
Special appliance load: electric range (see 220.19)	8000 VA

Minimum Number of Branch Circuits Required for Each Dwelling Unit [See 210.11(A)]:

General lighting load: 2520 volt-amperes ÷ 120 V = 21 amperes or two 15-ampere, 2-wire circuits; or two 20-ampere, 2-wire circuits.

Small appliance load: Two 2-wire circuits of 12 AWG wire [see 220.11(C)(1)].

Range circuit: 8000 volt-amperes ÷ 240 volts = 33 amperes or a circuit of two 8 AWG conductors and one 10 AWG conductor as permitted by 220.22. [See 210.19(C).]

Minimum Size Feeder Required for Each Dwelling Unit (See 215.2)

Computed load (Article 220)	
General lighting	2520 VA
Small appliance (two 20-ampere circuits)	3000 VA
Subtotal computed load (without ranges)	5520 VA

Application of Demand Factor (see Table 220.11)

First 3000 volt-amperes at 100%	3000 VA
5520 VA – 3000 VA = 2520 volt-amperes at 35%	882 VA
Net computed load (without ranges)	3882 VA
Range load	8000 VA
Net computed load (with ranges)	11,882 VA

Size of Each Feeder (See 215.2)

For 120/240-volt, 3-wire system (without ranges):

Net computed load: 3882 VA ÷ 240 V = 16.2 A

For 120/240-volt, 3-wire system (with ranges):

Net computed load: 11,882 VA ÷ 240 V = 49.5 A

Feeder Neutral

Lighting and small appliance	3882 VA
Range: 8000 volt-amperes at 70% (see Section 220.22) (only for apartments with electric ranges)	5600 VA
Net computed load (neutral)	9482 VA

Calculated Load for Neutral

9482 VA ÷ 240 V = 39.5 A

Minimum Size Feeders Required from Service Equipment to Meter Bank (for 20 Dwelling Units—10 with Ranges)

Total computed load:	
Lighting and small appliance 20 × 5520 volt-amperes	110,400 VA
Application of Demand Factor:	
First 3000 volt-amperes at 100%	3000 VA
110,400 VA − 3000 VA = 107,400 volt-amperes at 35%	37,950 VA
Net computed load	40,590 VA
Range: 10 ranges (less than 12 kVA), (see Col. A, Table 220.19)	25,000 VA
Net computed load (with ranges)	65,590 VA

Net computed load for 120/240-volt, 3-wire system:

65,590 VA ÷ 240 V = 273 A

Feeder Neutral

Lighting and small appliance	40,590 VA
Range: 25,000 volt-amperes at 70% (see 220.22)	17,500 VA
Net computed load (neutral)	58,090 VA

Calculated Load for Neutral

58,090 VA ÷ 240 V = 242 A

Further Demand Factor (220.22):

200 amperes at 100%	200 A
242-200 = 42 amperes at 70%	29 A
Net computed load (neutral)	229 A

Minimum Size Main Feeder (or Service Conductors) Required (Less House Load) (for 40 Dwelling Units—20 with Ranges)

Total computed load:	
Lighting and small appliance load,	
40 × 5520 volt-amperes	220,800 VA
Application of Demand Factor (from Table 220.11)	
First 3000 volt-amperes at 100%	3000 VA
Next 120,000 VA – 3000 VA =	
117,000 volt-amperes at 35%	40,950 VA
Remainder 220,800 VA – 120,000 VA =	
100,800 volt-amperes at 25%	25,200 VA
Net computed load	69,150 VA
Range load, 20 ranges (less than	
12 kVA) (see Col. C, Table 220.19)	35,000 VA
Net computed load	104,150 VA

For 120/240-volt, 3-wire system:

Net computed load: 104,150 VA ÷ 240 V = 434 A

Feeder Neutral

Lighting and small appliance load	69,150 VA
Range load, 35,000 volt-amperes at 70%	
(see 220.22)	24,500 VA
Computed load (neutral)	93,650 VA

93,650 VA ÷ 240 V = 390 A

Further Demand Factor (See 220.22)

200 amperes at 100%	200 A
390A – 200 A = 190 amperes at 70%	133 A
Net computed load (neutral)	333 A

[See Tables 310.16 through 310.21 and 310.15(B)(2) and (B)(4).]

Example No. D4(b) Optional Calculation for Multifamily Dwelling

A multifamily dwelling equipped with electric cooking and space heating or air conditioning and having 40 dwelling units.

Meters in two banks of 20 each plus house metering and individual subfeeders to each dwelling unit.

Each dwelling unit is equipped with an electric range of

8-kW nameplate rating, four 1.5-kW separately controlled 240-volt electric space heaters, and a 2.5-kW 240-volt electric water heater. Assume range, space heater, and water heater kW ratings equivalent to kVA.

A common laundry facility is available to all tenants [see 210.52(F), Exception No. 1].

Area of each dwelling unit is 840 square feet.

Computed Load for Each Dwelling Unit (See Article 220)

General lighting load:	
840 sq. ft. at 3 volt-amperes per sq. ft.	2520 VA
Electric range	8000 VA
Electric heat, 6 kVA (or air conditioning if larger)	6000 VA
Electric water heater	2500 VA

Minimum Number of Branch Circuits Required for Each Dwelling Unit

General lighting: 2520 volt-amperes ÷ 120 volts = 21 amperes. Two 15-ampere 2-wire circuits or two 20-ampere 2-wire circuits.

Small appliance load: Two 2-wire circuits of 12 AWG [see 220.11(C)(1)].

Range circuit: 8000 volt-amperes × 80% ÷ 240 volts = 27 amperes on a circuit of three 10 AWG conductors as permitted in Column B of Table 220.19.

Space heating: 6000 volt-amperes ÷ 240 volts = 25 amperes

No. of circuits (see 210.11).

Minimum Size Feeder Required for Each Dwelling Unit (See 215.2)

Computed load (see Article 220)	
General lighting load	2520 VA
Small appliance load, two 20-ampere circuits	3000 VA
Subtotal computed load (without range and space heating)	5520 VA
Application of Demand Factor	
First 3000 volt-amperes at 100%	3000 VA
5520 VA – 3000 VA = 2520 volt-amperes at 35%	882 VA
Net computed load (without range and space heating)	3882 VA

Range	6400 VA
Space heating (see 220.15)	6000 VA
Water heater	2500 VA
Net computed load (for individual dwelling unit)	18,782 VA

Size of Each Feeder

For 120/240 volt, 3-wire system:

Net computed load of 18,782 VA ÷ 240 V = 78 A

Feeder Neutral (See 220.22)

Lighting and small appliance load	3882 VA
Range 6400 at 70% (see 220.22)	4480 VA
Space and water heating (no neutral) 240 volt	0 VA
Net computed load (neutral)	8362 VA

Calculated Load for Neutral

8362 VA ÷ 240 V = 35 A

Minimum Size Feeder Required from Service Equipment to Meter Bank (for 20 Dwelling Units)

Total computed load:	
Lighting and small appliance load, 20 units × 5520 volt-amperes	110,400 VA
Water and space heating load, 20 units × 8500 volt-amperes	170,000 VA
Range load, 20 × 8000	160,000 VA
Net computed load (20 dwelling units)	440,400 VA
Net computed load using optional calculation (see Table 220.32) 440,400 volt-amperes × 0.38	167,352 VA

167,352 VA ÷ 240 V = 697 A

Minimum Size Main Feeder Required (Less House Load) (for 40 Dwelling Units)

Computed load	
Lighting and small appliance load, 40 units × 5520 volt-amperes	220,800 VA

Water and space heating, 40 units × 8500 volt-amperes	340,000 VA
Range: 40 ranges × 8000 volt-amperes	320,000 VA
Net computed load (40 dwelling units)	880,800 VA
Net computed load using optional calculation (See Table 220.32) 880,800 volt-amperes × 0.28 =	246,624 VA

246,624 VA ÷ 240 V = 1028 A

Feeder Neutral Load for Feeder from Service Equipment to Meter Bank (for 20 Dwelling Units)

Lighting and small appliance load, 20 units × 5520 volt-amperes	110,400 VA
First 3000 volt-amperes at 100%	3,000 VA
110,000 VA – 3000 VA = 107,400 volt-amperes at 35%	37,590 VA
Net computed load	40,590 VA
20 ranges: 35,000 volt-amperes at 70% (see Table 220.19 and 220.22)	24,500 VA
Total	65,090 VA

65,090 VA ÷ 240 V = 271 A

Further Demand Factor (See 220.22)

First 200 amperes at 100%	200 A
Balance: 271 A – 200 A = 71 amperes at 70%	50 A
Total	250 A

Feeder Neutral Load of Mains Feeder (Less House Load) (for 40 Dwelling Units)

Lighting and small appliance load, 40 × 5520 volt-amperes	220,800 VA
First 3000 volt-amperes at 100%	3,000 VA
Next 120,000 volt-amperes – 3000 volt-amperes = 117,000 volt-amperes at 35%	40,950 VA
Remainder 220,800 volt-amperes – 120,000 volt-amperes = 100,800 volt-amperes at 25%	25,200 VA
Net computed load	69,150 VA

40 ranges: 55,000 volt-amperes at 70% (see Table 220.19 and 220.22)	38,500 VA
Total	107,650 VA

107,650 VA ÷ 240 V = 449 A

Further demand factor (see 220.22)

First 200 amperes at 100%	200 A
Balance: 449 A − 200 A = 249 amperes at 70%	174 A
Total	374 A

Example No. D5(a) Multifamily Dwelling Served at 208Y/120 Volts, Three Phase

All conditions and calculations are the same as for the multifamily dwelling of Example No. D4(a) served 120/240 volts, single phase except service to each dwelling unit would be two-phase legs and neutral.

Minimum Number of Branch Circuits Required for Each Dwelling Unit (See 220.11)

Range circuit: 8000 volt-amperes ÷ 208 volts = 38 amperes or a circuit of two 8 AWG conductors and one 10 AWG conductor as permitted by 210.19 Column B.

Minimum Size Feeder Required for Each Dwelling Unit (See 215.2)

For 120/208-volt, 3-wire system (without ranges)

Net Computed Load of 3882 volt-amperes ÷ 2 legs ÷ 120 volts/leg = 16.2 amperes

For 120-208-volt, 3-wire system (with ranges)

Net computed load (range) of 8000 volt-amperes ÷ 208 volts = 38.5 amperes
Total load (range plus lighting) = 38.5 VA + 16.2 A = 54.7 amperes
Feeder neutral: (range) of 8000 VA × 70% = 5600 volt-amperes ÷ 208 volts = 26.9 amperes
Total load (range plus lighting) = 26.9 A + 16.2 A = 43.1 A

Minimum Size Feeders Required from Service Equipment to Meter Bank (for 20 Dwelling Units—10 with Ranges)

For 208Y/120-volt, 3-phase, 4-wire system:

Ranges: maximum number between any two phase legs = 4. 2 × 4 = 8. Table 220.19 Demand = 23,000 volt-amperes
Per phase demand: 23,000 volt-amperes ÷ 2 = 11,500 volt-amperes
Equivalent 3-phase load = 34,500 volt-amperes
Net computed load (total): 40,590 volt-amperes + 34,500 volt-amperes = 75,090 volt-amperes

$$75{,}090 \text{ VA} \div (208)(1.732) = 208.4 \text{ A}$$

Feeder Neutral Size

Net computed lighting and appliance load and equivalent range load

$$40{,}590 \text{ VA} + (34{,}500 \text{ VA} \times 70\%) = 64{,}700 \text{ VA}$$

Net computed neutral load:

$$64{,}700 \text{ VA} \div (208)(1.732) = 179.7 \text{ A}$$

Minimum Size Main Feeder (Less House Load) (for 40 Dwelling Units—20 with Ranges)

For 208Y/120-volt, 3-phase, 4-wire system:

Ranges: maximum number between any two phase legs = 7

2 × 7 = 14.
Table 220.19 Demand = 29,000 volt-amperes
Per phase demand: 29,000 VA ÷ 2 = 14,500 volt-amperes
Equivalent 3-phase load = 43,500 volt-amperes
Net computed lighting and appliance load and equivalent range load: 69,150 volt-amperes + 43,500 volt-amperes = 112,650 volt-amperes

$$112{,}650 \text{ VA} \div (208)(1.732) = 312.7 \text{ A}$$

Main Feeder Neutral Size

$$69{,}150 \text{ VA} + (43{,}500 \text{ VA} \times 70\%) = 99{,}600 \text{ VA}$$

99,600 volt-amperes ÷ (208)(1.732) = 276.5 amperes

Further Demand Factor (See 220.22)

200 amperes at 100%	200.0 A
276.5 – 200 A = 76.5 amperes at 70%	53.6 A
Net computed load neutral	253.6 A

Example No. D5(b) Optional Calculation for Multifamily Dwelling Served at 208Y/120 Volts, Three Phase

All conditions and calculations the same as for Optional Calculations for Multifamily Dwelling of Example No. D4(b) served at 120/40 volt, single phase except service to each dwelling unit shall be two-phase legs and neutral.

Minimum Number of Branch Circuits Required for Each Dwelling Unit (See 210.11)

Range circuit: 8000 volt-amperes × 80% ÷ 208 volt = 30.7 amperes or a circuit of two 8 AWG conductors and one 10 AWG conductor as permitted by 220.19(B).

Space heating: 6000 volt-amperes ÷ 208 volts = 28.8 amperes. Two 20-ampere, 2-pole circuits required, 12 AWG conductors.

Minimum Size Feeder Required for Each Dwelling Unit

120/208-volt, 3-wire circuit

Net computed load: 18,782 volt-amperes ÷ 208 volts	= 90.3 amperes
Net computed lighting load (line to neutral): 3882 volt-amperes ÷ 2 legs ÷ 120 volts/leg	= 16.2 amperes
Line to line = 14,900 volt-amperes ÷ 208 volts	71.6 amperes
Total load = 16.2 A + 71.6 A =	87.8 amperes

Minimum Size Feeder Required for Service Equipment to Meter Bank (20 Dwelling Units)

Net Computed Load

167,352 volt-amperes ÷ (208)(1.732) = 464.9 amperes.

Feeder Neutral Load

65,080 VA ÷ (208)(1.732) = 180.65 A

Minimum Size Main Feeder Required (Less House Load) (40 Dwelling Units)

Net Computed Load

246,624 volt-amperes ÷ (208 V)(1.732) = 684.6 amperes

Main Feeder Neutral Load

107,650 volt-amperes ÷ (208 V)(1.732) = 298.8 amperes

Further Demand Factor (See 220.22)

200 amperes at 100%	200.0 A
298A – 200A = 98.8 amperes at 70%	69.2 A
Net computed load (neutral)	269.2 A

Example No. D6 Maximum Demand for Range Loads

Table 220.19, Column C, applies to ranges not over 12 kW. The application of Note 1 to ranges over 12 kW (and not over 27 kW) and Note 2 to ranges over 8 3/4 kW (and not over 27 kW) is illustrated in the following examples:

A. Ranges all the same rating (See Table 220.19, Note 1)

Assume 24 ranges each rated 16 kW. From Table 220.19 Column C the maximum demand for 24 ranges of 12 kW rating is 39 kW.

16 kW exeeeds 12 kW by 4
5% × 4 = 20% (5% increase for each kW in excess of 12)
39 kW × 20% = 7.8 kW increase
39 kW + 7.8 kW = 46.8 kW (value to be used in selection of feeders)

B. Ranges of unequal rating (See Table 220.19, Note 2)

Assume:

5 ranges each rated 11 kW
2 ranges each rated 12 kW
20 ranges each rated 13.5 kW
3 ranges each rated 18 kW

5 × 12 = 60 kW (Use 12 kW for range rated less than 12)
2 × 12 = 24 kW
20 × 13.5 = 270 kW
3 × 18 = 54 kW
30 ranges 408 kW total kW

408 kW ÷ 30 ranges = 13.6 kW (average to be used for computation)

From Table 220.19, Column C, the demand for 30 ranges of 12 kW rating is 15 kW + (1 kW × 30 ranges) = 45 kW.

13.6 exceeds 12 by 1.6 (use 2 kW)
5% × 2 = 10% (5% increase for each kW in excess of 12 kW)
45 kW × 10% = 4.5 kW increase
45 kW + 4.5 kW = 49.5 kW: (value to be used in selection of feeders)

Example No. D8 Motors, Conductors, Overload, and Short-Circuit and Ground-Fault Protection

(See 240.6, 430.6, 430.22, 430.23, 430.24, 430.32, 430.52, and 430.62, Tables 430.52 and 430.150.)

Determine the minimum required conductor ampacity, the motor overload protection, the branch-circuit short-circuit and ground-fault protection, and the feeder protection, for three induction-type motors on a 480-V, 3-phase feeder, as follows:

(a) One 25-hp, 460-V, 3-phase, squirrel-cage motor, nameplate full-load current 32 A, Design B, Service Factor 1.15

(b) Two 30-hp, 460-V, 3-phase, wound-rotor motors, nameplate primary full-load current 38 A, nameplate secondary full-load current 65 A, 40°C rise.

Conductor Ampacity

The full-load current value used to determine the minimum required conductor ampacity is obtained from Table 430.150 [see 430.6(A)] for the squirrel-cage motor and the primary of the wound-rotor motors. To obtain the minimum required conductor ampacity, the full-load current is multiplied by 1.25 [see 430.22 and 430.23(A)].

For the 25-hp motor, 34 A × 1.25 = 42.5 A
For the 30-horsepower motors, 40 A × 1.25 = 50 A
65 A × 1.25 = 81.25 A

Motor Overload Protection

Where protected by a separate overload device, the motors are required to have overload protection rated or set to trip at not more than 125% of the nameplate full-load current [see 430.6(A) and 430.32(A)(1)].

For the 25-hp motor, 32 A × 1.25 = 40 A

For the 30-hp motors, 38 A × 1.25 = 47.5 A

Where the separate overload device is an overload relay (not a fuse or circuit breaker), and the overload device selected at 125% is not sufficient to start the motor or carry the load, the trip setting is permitted to be increased in accordance with 430.32(C).

Branch-Circuit Short-Circuit and Ground-Fault Protection

The selection of the rating of the protective device depends on the type of protective device selected, in accordance with 430.52 and Table 430.152. The following is for the 25-hp motor.

(a) Nontime-Delay Fuse. The fuse rating is 300% × 34 A = 102 A. The next larger standard fuse is 100 A [see 240.6 and 430.52(C)(1), Exception No. 1]. If the motor will not start with a 110-A nontime-delay fuse, the fuse rating is permitted to be increased to 125 A because this rating does not exceed 400% [see 430.52(C)(1), Exception No. 2(a)]

(b) Time-Delay Fuse. The fuse rating is 175% × 34 A = 59.5 A. The next largest fuse is 60-A fuse [see 240.6 and 430.52(C)(1), Exception No. 1]. If the motor will not start with a 60-A time-delay fuse, the fuse rating is permitted to be increased to 70 A because this rating does not exceed 225% [see 430.52(C)(1), Exception No. 2b].

Feeder Short-Circuit and Ground-Fault Protection

The rating of the feeder protective device is based on the sum of the largest branch-circuit protective device (example is 110 A) plus the sum of the full-load currents of the other motors, or 110 + 40 + 40 = 190 A. The nearest standard fuse that does not exceed this value is 175 A [see 240.6 and 430.62(A)].

Example No. D9 Feeder Ampacity Determination for Generator Field Control

(See 215.2; 430.24 Exception No. 1; 620.13, 620.14, 620.61, and Tables 430.22(B) and 620.14.)

Determine the conductor ampacity for a 460-volt 3-phase, 60-Hz-ac feeder supplying a group of six elevators.

The 460-volt-ac drive motor nameplate rating of the largest

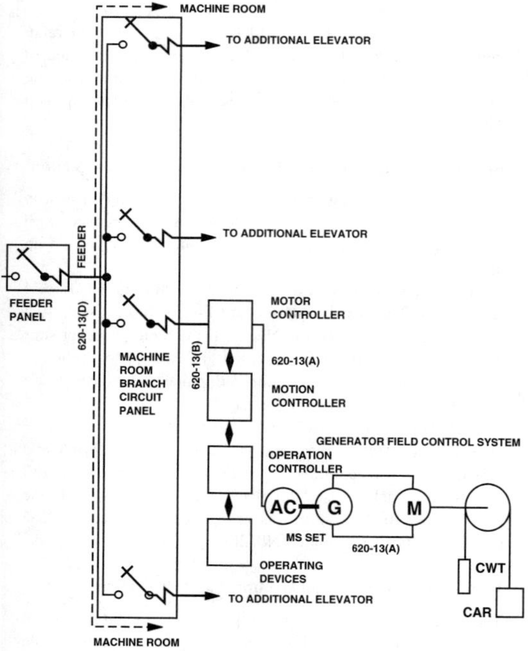

Figure D9
Generator field control

MG set for one elevator is 40 horsepower, and 52 amperes, and the remaining elevators, each have a 30-horsepower, 40-ampere, ac drive motor rating for their MG sets. In addition to a motor controller, each elevator has a separate motion/operation controller rated 10 amperes continuous to operate microprocessors, relays, power supplies, and the elevator car door operator. The MG sets are rated continuous.

Conductor Ampacity

Conductor ampacity is determined as follows:

(a) Per 620.13(D) and 620.61(B)(1) use Table 430.22(B), for intermittent duty (elevators). For intermittent duty using a continuous rated motor, the percentage of nameplate current rating to be used is 140 percent.

(b) For the 30-horsepower ac drive motors, 140% × 40 amps = 56 amperes.

(c) For the 40-horsepower ac drive motors, 140% × 52 amps = 73 amperes.

(d) The total conductor ampacity is the sum of all motor currents (1 motor × 73 amperes) + (5 motors × 56 amperes) = 353 amperes.

(e) Per 620.14 and Table 620.14, the conductor (feeder) ampacity shall be permitted to be reduced by the use of a demand factor. Constant loads are not included. (See 620.14, FPN). For six elevators, the demand factor is 0.79. Therefore feeder diverse ampacity is, 0.79 × 353 amperes = 279 amperes.

(f) Per 430.24 and 215.3, controller continuous current is 125% × 10 amperes = 12.5 amperes.

(g) The total feeder ampacity is the sum of the diverse current and all the controller continuous current. I total = 279 + (6 elevators × 12.5) = 354 amperes.

(h) This ampacity shall be permitted to be used to select the wire size.

See Figure D9.

Example No. D10 Feeder Ampacity Determination for Adjustable-Speed Drive Control [See 215.2; 430.24; 430.24(B) 620.13, 620.14, 620.61 Tables 430.22(B), and 620.14]

Determine the conductor ampacity for a 460-volt, 3-phase, 60 Hz ac feeder supplying a group of six identical elevators. The system is adjustable-speed SCR dc drive. The power transformers are external to the drive (motor controller) cabinet. Each elevator has a separate motion/operation controller connected to

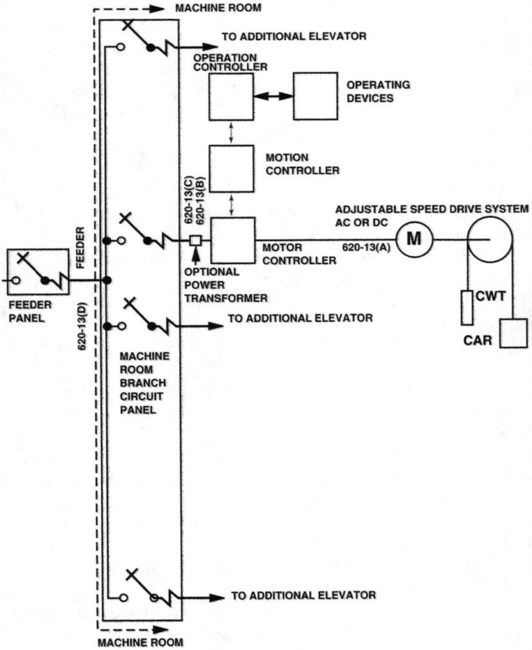

Figure D10
Adjustable speed drive control

the load side of the main line disconnect switch rated 10 amperes continuous to operate microprocessors, relays, power supplies, and the elevator car door operator. Each transformer is rated 95 kVA with an efficiency of 90 percent.

Conductor Ampacity

Conductor ampacity is determined as follows:

(a) Calculate the nameplate rating of the transformer:

$$I = \frac{95 \text{ kVA} \times 1000}{\sqrt{3} \times 460 \text{ V.} \times 0.90 \text{ eff.}} = 133 \text{ amperes}$$

(b) Per 620.13(D), for six elevators, the total conductor ampacity is the sum of all the currents: 6 elevators × 133 amperes = 798 amperes.

(c) Per 620.14 and Table 620.14, the conductor (feeder) ampacity would be permitted to be reduced by the use of a demand factor. Constant loads are not included. (See 620.13, FPN No. 2). For six elevators, the demand factor is 0.79. Therefore feeder diverse ampacity = 0.79 × 798 amps = 630 amperes.

(d) Per 430.24 and 215.3, the controller continuous current is 125% × 10 amperes = 12.5 amperes.

(e) The total feeder ampacity is the sum of the diverse current and all the controller constant current. I total = 630 + (6 elevators × 12.5) = 705 amperes.

(f) This ampacity would be permitted to be used to select the wire size.

See Figure D10.

Example No. D11 Mobile Home

A mobile home floor is 70 ft. by 10 ft. and has two small appliance circuits; a 1000-VA, 240-V heater; a 200-VA, 120-V exhaust fan; a 400-VA, 120-V dishwasher; and a 7000-VA electric range.

Lighting and Small Appliance Load

Lighting (70 ft. × 10 ft. × 3 VA per ft.^{2})	2100 VA
Small appliance (1500 VA × 2 circuits)	3000 VA
Laundry (1500 VA × 1 circuit)	1500 VA
	6600 VA
First 3000 VA at 100%	3000 VA
Remainder (6600 VA – 3000 VA) at 35%	1260 VA
Total	4260 VA

4260 VA ÷ 240 V = 17.75 A per leg

Amperes per Leg

	Leg A	Leg B
Lighting and appliances	17.75	17.75
Heater (1000 VA ÷ 240 V)	4.20	4.20
Fan (200 VA × 125% ÷ 120 V)	2.08	—
Dishwasher (400 VA ÷ 120 V)	—	3.30
Range (7000 VA × 0.8 ÷ 240 V)	23.30	23.30
Total amperes per leg	47.33	48.55

Based on the higher current calculated for either leg, a minimum 50-A supply cord would be required.

For SI units, 0.093 m^2 = 1 ft^2 and 0.3048 m = 1 ft.

Example No. D12 Park Trailer

A park trailer floor is 40 ft. by 10 ft. and has two small appliance circuits; a 1000-VA, 240-V heater; a 200-VA, 120-V exhaust fan; a 400-VA, 120-V dishwasher; and a 7000-VA electric range.

Lighting and Small Appliance Load

Lighting (40 × 10 × 3 VA per ft.^{2})	1200 VA
Small appliance (1500 VA × 2 circuits)	3000 VA
Laundry (1500 VA × 1 circuit)	1500 VA
Subtotal	5700 VA
First 3000 VA at 100%	3000 VA
Remainder (5700 VA – 3000 VA = 2700 VA) at 35%	945 VA
Total	3945 VA

3945 VA ÷ 240 V = 16.44 A per leg

Amperes per Leg

	Leg A	Leg B
Lighting and appliances	16.44	16.44
Heater (1000 VA ÷ 240 V)	4.20	4.20
Fan (200 VA × 125% ÷ 120 V)	2.08	—
Dishwasher (400 VA ÷ 120 V)	—	3.3
Range (7000 VA × 0.8 ÷ 240 V)	23.30	23.30
Total amperes per leg	46.02	47.24

Based on the higher current calculated for either leg, a minimum 50-A supply cord would be required.

For SI units, 0.093 m^2 = 1 ft^2 and 0.3048 m = 1 ft.

Annex E

TYPES OF CONSTRUCTION

Table E.1 Fire Resistance Ratings (in hours) for Type I through Type V Construction

	Type I		*Type II*			*Type III*		*Type IV*	*Type V*	
	443	***332***	***222***	***111***	***000***	***211***	***200***	***2HH***	***111***	***000***
Exterior Bearing Walls—										
Supporting more than one floor, columns, or other bearing walls. . . .	4	3	2	1	0[1]	2	2	2	1	0[1]
Supporting one floor only.	4	3	2	1	0[1]	2	2	2	1	0[1]
Supporting a roof only.	4	3	1	1	0[1]	2	2	2	1	0[1]
Interior Bearing Walls—										
Supporting more than one floor, columns, or other bearing walls. . . .	4	3	2	1	0	1	0	2	1	0
Supporting one floor only.	3	2	2	1	0	1	0	1	1	0
Supporting roofs only.	3	2	1	1	0	1	0	1	1	0
Columns—										
Supporting more than one floor, columns, or other bearing walls. . . .	4	3	2	1	0	1	0	H[2]	1	0
Supporting one floor only.	3	2	2	1	0	1	0	H[2]	1	0
Supporting roofs only.	3	2	1	1	0	1	0	H[2]	1	0

Table E.1 Fire Resistance Ratings (in hours) for Type I through Type V Construction (continued)

	Type I		*Type II*			*Type III*		*Type IV*	*Type V*	
	443	***332***	***222***	***111***	***000***	***211***	***200***	***2HH***	***111***	***000***
Beams, Girders, Trusses & Arches—										
Supporting more than one floor, columns, or other bearing walls. . . .	4	3	2	1	0	1	0	H[2]	1	0
Supporting one floor only.	3	2	2	1	0		0	H[2]	1	0
Supporting roofs only.	3	2	1	1	0	1	0	H[2]	1	0
Floor Construction	3	2	2	1	0	1	0	H[2]	1	0
Roof Construction	2	1½	1	1	0	1	0	H[2]	1	0
Exterior Nonbearing Walls[3]	0[1]	0[1]	0[1]	0[1]	0[1]	0[1]	0[1]	0[1]	0[1]	0[1]

(Shaded) These members that shall be permitted to be of approved combustible material.

Source: Table 3-1 from NFPA 220, *Standard on Building Construction,* 1999.

[1]See A-3 1 in NFPA 220.

[2]"H" indicates heavy timber members; see text for requirements.

[3]Exterior nonbearing walls meeting the conditions of acceptance of NFPA 285, *Standard Method of Test for the Evaluation of Flammability Characteristics of Exterior Non-Load-Bearing Wall Assemblies Containing Combustible Components Using the Intermediate-Scale, Multistory Test Apparatus,* shall be permitted to be used.